FAR FROM HOME

A NOVEL OF THE WAR IN IRAQ

JEREMIAH COE AND ADAM D. BLOCHARD

HOPKIN'S HOUSE
PUBLISHING

Far From Home : A Novel of the War in Iraq

ISBN: 978-1626769045

Copyright © 2016 by Jeremiah Coe & Adam D. Blochard

Library of Congress CIP data applied for.

First Edition: July 2016

Other works by Jeremiah Coe

NOVELS :

The Dead of Space Book One: Brave New World

The Dead of Space Book Two: Journey's End

Here Comes Santa

Uncivil Dead

E-BOOKS :

Vampire's Retribution

Love Never Dies

Justice

Three Wishes

SHORT STORY COLLECTIONS :

Tales From A Twisted Mind Volume One

A NOTE FROM JEREMIAH

I have to admit, as I sit down to write this novel, that this is the first time I've been nervous about starting a new project. There are two reasons for this.

First, even though this is a work of fiction, the Battle of An Nasiriyah, where many American Marines and soldiers, including my co-author, fought, was used as the template for the fictional battle in *Far From Home*. For those of us who have never been in combat, we see military movies and books as nothing more than a source of entertainment. What we forget is that for many of our fellow Americans, it is real life, not just a work of fiction. I don't want to write this story just to write one of the first, if not *the* first, fictional novel about the War in Iraq. Job one for me in writing this novel is to do right by the men and women who have fought in Iraq and Afghanistan. At the same time, I want to write a novel that is apolitical, I don't want this to carry a pro-war or anti-war message because, regardless of my personal views, I will always support the men and women who go when our nation's leadership says go.

Second, Adam and I will not be keeping any of the author royalties that are earned from the sale of this novel. Every penny of our author royalties will go to Operation Ward 57, which helps those who have returned from service in Iraq or Afghanistan with life-altering injuries while they are being cared at Walter Reed National Military Medical Center. So I hope to make this a book that you'll suggest to others because it's not us, the authors, who are receiving the benefit of each purchase, it is the men and women who went to war and came back injured who are. I would encourage you to visit the Operation Ward 57 website at http://www.operationward57.org. If you enjoy reading this story, please consider donating a copy to your local library. It might sound strange, but I really do hope this novel outsells the novels that I do receive a royalty from.

Since this is a work of fiction, it should not be taken as any kind of historical record. For nonfiction accounts of the Battle of An Nasiriyah, I would suggest reading *An Nasiriyah: The Fight for the*

Bridges by Gary Livingston and *Marines in the Garden of Eden* by Richard S. Lowery.

I wish I could thank everyone who has ever served in the U.S. military by name, but there is just no way that I can do that. So, I'd like to thank everyone who has worn the uniforms of the U.S. Army, the U.S. Marine Corps, the U.S. Air Force or the U.S. Navy from the time of the Revolutionary War to today, regardless of whether your service was in peacetime or wartime. Thank you.

CHAPTER ONE

Private First Class Ron Farber manned the fifty-caliber machine gun on top of a Humvee, which was part of the massive land armada moving across the desert sea of Iraq. In June of 2002, he had been excited and full of hope for the future while graduating from high school in Bowling Green, Ohio, and now, in March of 2003, he was part of the first infantry unit pushing into Iraq.

At the age of nineteen, he was too young to remember America throwing Saddam Hussein's army out of Kuwait in Operation Desert Storm, but he would always remember America coming to Iraq to throw Saddam Hussein's government out of power in Operation Iraqi Freedom because he was part of it. He was proud to be here. He was proud to be a United States Marine.

It had been two days since his battalion left Camp Nicholas, the tent city in Kuwait. Camp Nicholas had been named after Samuel Nicholas, who led the Marine Corps from 1775 to 1781. He was widely considered to have been the first Commandant of the Marine Corps, even though he led it before the title existed.

Farber had been raised on a steady diet of movies like *Heartbreak Ridge*, *Full Metal Jacket*, *Platoon* and *Hamburger Hill* from a young age, and, like most young men who've gone off to war without a real understanding of what war is; he thought he was in for an exciting adventure. He had been eager to go to war and, even though they would be at their objective and fighting in only a couple of hours, it seemed to be taking forever. He was a Marine. Marines fight, and fight is exactly what Ron Farber wanted to do.

He had wanted to be a U.S. Marine since he was five years old and saw his first Marine in his blue dress uniform. While most little boys dreamed of being a cowboy, an astronaut or a firefighter, Farber had never wanted anything more than to carry an M-16 as a Marine Corps infantryman. He had never really had an interest in playing sports, but in high school he had spent four years on the football and wrestling teams for the sole purpose of preparing himself physically for recruit training. And now he was here in Bravo Company, he was doing what it is that Marines were bred to do and he couldn't imagine being anywhere else. Despite the discomfort of being partially dressed in his Nuclear Biological and Chemical (NBC) suit in such heat, Ron Farber was happy.

He looked at the NBC suit as a necessary inconvenience. They were here to rid Iraq not only of Saddam Hussein, but also of his weapons of mass destruction. In the past, Saddam had turned those weapons on his own people, so it would have been foolhardy to assume he would not likewise turn them on the U.S. forces that had come to remove him, and his entire government, from power. There had been countless drills on getting into the NBC suits quickly while they were in Camp Nicholas and, all of his complaining about the drills aside, Ron was happy to have had them now that they had crossed the border between Kuwait and Iraq, now that the war had started for him. Ron had gotten good at getting into his suit quickly.

The thought of going to college as soon as he graduated high school had never occurred to him. The day after graduation, he and two of his friends drove to the recruiters' office and enlisted. Ron hadn't ruled out the possibility of going to college though. As a matter of fact, he had been looking at schools with Naval ROTC programs so he could continue his career as a Marine after college, only as an officer instead of an enlisted man. He had even thought that he might make a good Midshipman at the U.S. Naval Academy at Annapolis. Ron didn't know what his future held for sure, but he did know that he couldn't imagine a future where he wasn't wearing the Eagle, Globe and Anchor.

Ron knew, more in an abstract sense considering that he had not actually seen battle yet, that he could be killed, but at the same time he felt good about his chances of seeing his family again. After all, he was a Marine; he was part of the United States military, the best-trained and best-equipped fighting force in the world. For him, death was something that would come for him at some point in the far future, not today, not only after nineteen short years.

Looking around as the Humvee drove through the desert, Ron couldn't see how anyone could live in such a place. It was flat, it was dry, and, besides the occasional palm tree, Ron hadn't seen anything but what appeared to be endless sand. He couldn't fathom why anyone would want to live in such a God forsaken place. He certainly wouldn't have wanted to.

CHAPTER TWO

Alpha Company Sergeant Tom Daley, twenty-six-years old, from Pioche, Nevada, didn't have the same nervous excitement that many of his brother Marines around him in the AAV, which were also known as tracks, had. He was just all-around nervous. A nagging little voice in the back of his head kept telling Daley, "Today is the day you're going to die," and, try as he might, he couldn't get the voice to shut up. Unlike most of the Marines around him, Daley had joined the Corps to escape small town life. Pioche, Nevada, had a population of around nine hundred people, three bars, one bank, a museum, a library and a couple of little shops. It was the type of place where if you farted, everyone knew about it immediately. Not to speak badly of Pioche. It is very neighborly, something that most places in America can no longer say, and if someone needs any kind of help, everyone else is right there helping them.

Daley hadn't hated living in Pioche; quite to the contrary, he had loved everything about it. He had just wanted to see, and do, more. There was a really strong possibility that he would eventually return to Pioche and settle down, but not until he had seen all he wanted to see and done all that he wanted to do.

Like most teenage boys, Daley had been in more than his fair share of fistfights and he quickly developed the reputation of being one not to cross. Even though he lacked any formal training in wrestling, boxing or the martial arts, Daley discovered that he enjoyed fighting and decided he would like to find a way to do it without getting himself into trouble.

Having grown up on a steady diet of military action movies, it seemed perfectly reasonable to Daley that he should find a home after high school in the military. He spent time talking to the recruiters from all four branches and gave serious thought to what they all said prior to signing with any of them. Tom ruled out the Navy because there was just something about fighting on the sea that did not appeal to him. If he was to die, he didn't want to drown or be eaten by sharks. He had likewise ruled out the Air Force because most of the jobs seemed to be noncombatant and that wasn't what he wanted either. So it was down to a choice between the Army and the Marine Corps. He had always enjoyed testing himself, so the longer, and apparently more intensive, recruit training of the Marine Corps appealed more to him than the Army did. And so, here he was in Iraq as a Marine.

Now, as Tom Daley sat in an AAV with eleven other Marines, heading towards a city named El Jasiph, where the first infantry battle of the war would occur, the very real possibility that he could die at the age of twenty-six hit home and he wasn't so sure he hadn't made a mistake. The Air Force and the Navy were starting to look much better to him with each mile of Iraqi desert that passed.

The last thing Daley wanted was to let those around him see his nervousness. He was certain it would cause him to become a subject of their ridicule and scorn. In the best case, he figured he would be the butt of their jokes for quite some time to come. They were supposed to have left all personal effects back at Camp Nicholas, but Daley had brought a well-read copy of the novel *Vision Quest* by Terry Davis. It was one of his favorite books—he was on his third reading of the novel. For most of their trip from Camp Nicholas to El Jasiph, Daley had had his face buried in the book and did his best to avoid engaging in any of the banter that was going on around him. Little did he know that his reading and avoidance of conversation, had told the Marines around him everything they needed to know about his nervousness. Out of respect, they allowed him to keep to himself because a part of each of them felt the same way.

Some of the Marines in the battalion were eager for battle and chomping at the bit to get into their first firefight, but most of them had at least a seed of nervousness causing slight flutters in their stomachs. For Tom Daley, those flutters felt like an entire troupe of circus performers was in his stomach, all performing at the same time. Most of the Marines sought to alleviate their nervousness with bravado, joking banter and conversation, but Daley had turned inward to deal with his.

Many of the Marines around him had hopes of fighting well. Daley's biggest hope was not to urinate in his pants when the first bullet was fired at them.

CHAPTER THREE

In Charlie Company, Privates Burt Farrow, Andy Irvin and Manny Alverez, all 18 years old and fairly recently out of recruit training, were known as The Three Stooges because they all loved to goof around and make their fellow Marines laugh. The three of them became fast friends after being assigned to Charlie Company at Camp Lejeune, North Carolina. Even though they loved each other like brothers, the three of them had not found a single political topic they could see eye to eye on and their frequent political discussions often resulted in a friendly discussion held with raised voices. These discussions led quite a few of the others to call them The United Nations because Farrow was white, Irvin was black and Alverez was Hispanic. Usually being called The United Nations was enough to bring the discussion being held down in volume but not enough to end it completely.

Farrow wiped sweat from his forehead. "I hate these mobile fuckin' beer cans."

"Bumpy as hell ain't they," Irvin replied.

"Any idea how much longer until we get there?" Farrow asked, looking at both Irvin and Alverez.

Irvin shrugged.

"Are we there yet?" Alverez asked in a childish voice.

"Seriously, man, any idea? I'm just about sick of sitting around back here."

Alverez made an "I don't know" gesture with his hands and shook his head. "Guess we'll know as soon as we hear bullets zinging around outside of this beer can," he said.

A few minutes of silence passed between them while the other Marines in the track carried on their own conversations.

"Do you guys think it'll be bad when we get there?" Alverez asked.

"These bozos gave up awfully damn quick the last time we was here. I'd imagine they'll do the same this time too," Irvin answered.

"And if they don't?"

"We've got superior fire power. The Cobras have been pounding the shit out of them all day and we're coming in with tanks and CAATs. Any fuckin' Haji shooting at us is a dead fuckin' Haji," Farrow answered.

Alverez shook his head. "I don't mind telling you boys, I'm nervous as hell."

Irvin rested a reassuring hand on Alverez's shoulder. "We all are, brother, we all are, but the trick is not to let it control you. My old boxing coach used to say that fear is like a fire. If controlled, it can cook your food and heat your campsite. If uncontrolled, it'll burn down your house and kill your whole fuckin' family. Which kind of fear we have is up to each of us, so make your decision now, before we get there. Do you have a small, easily manageable campfire or a raging forest fire?"

"Yeah, that makes sense," Alverez said. Deep down, he worried about orphaning his two-year-old son. His son's mother had been killed in a car accident after he had deployed to Iraq. The court had awarded temporary custody of his son to his parents since the other set of grandparents were both drug addicts. He planned on bringing his son to live with him in North Carolina as soon as he could after finishing his tour of duty in Iraq.

Farrow ran his hand over Alverez's high and tight haircut, as someone would a child's shaggy and unruly hair. "Besides, honky, you're too ugly to die. Any Haji bullet coming at you will die of fright long before it has a chance of hitting your ugly ass."

Alverez's and Farrow's friendship had evolved quickly and since race didn't matter to either of them, Farrow had taken to calling Alverez names that were meant to be derogatory towards white people and Alverez returned the favor by calling him names that were meant to be derogatory towards Hispanics.

The three of them were like brothers, but Irvin had never shown any interest in participating in that kind of banter and, not knowing just how sensitive he was towards race, neither of them attempted to bring him into it. They figured if he wanted to join in, he would, but what they didn't know was that Irvin felt left out by their not including him and was simply waiting for one of them to make the first overture.

Turning his head to look at Farrow, Alverez smiled. "And compared to you wetback, I still look like a Greek God."

Alverez's retort caused all three of them to laugh. None of them had ever been in combat before, and they didn't have the slightest idea of what they were really in for, so they were all justifiably nervous. However, the camaraderie and the friendly back and forth made the not knowing easier. They weren't just friends, they were brothers and with brothers at your side, how bad could things really get?

CHAPTER FOUR

Forty-eight-year-old Lieutenant Colonel Jim Everett sat in the front passenger seat of the Battalion Command Humvee, puffing on a cigar to help steady his nerves, which seemed to grow worse and worse as El Jasiph came into view. Next to him, in the driver's seat, sat Lance Corporal Jack Tays, behind him sat Chief Warrant Officer Pete Lockhart, the Battalion Gunner, who was known by every Marine in the Battalion as, simply, Gunner." Next to Gunner sat Sergeant Major Art Jamison. Above them, manning the Humvee's fifty-caliber machine gun was Corporal Bruce Hawkins.

Everett knew it wouldn't do any good to give himself an ulcer over things he couldn't do anything about. His battalion had split up into three companies over an hour before: Bravo Company would be crossing the bridge on the west side of the city; Everett would be going with Alpha Company and entering the city over the eastern bridge; and Charlie Company would be going over the northern bridge and pushing straight through the center of El Jasiph. Charlie Company would be traveling the route where the heaviest resistance was expected. It would be Alpha and Bravo's responsibility to come in from the sides and cover Charlie's flanks. If things went according to plan, the three companies would meet up, reforming the entire battalion about three-quarters of the way through El Jasiph, and then finish pushing their way through to the southern end of the city together. Adding to his nerves was, just like most of the Marines under his command, the fact that this would be his first time in combat. Everett had been born in 1955 and was too young to have fought in Vietnam. When Desert Storm came around, he had been a Major and a Company Commander, he had been in the theater of operations, but the war had been so short his company had not seen combat.

If it came to a knock-down, drag-out, bare-knuckled brawl, the hope was that the plan would work flawlessly, that it would be fast and furious and then be over.

CHAPTER FIVE

Lance Corporal Gary Inglehart, of Bravo Company, stood with his finger on the trigger of his Humvee's fifty-caliber machine gun as the outskirts of El Jasiph came into sight. His stomach felt like it did when he was on the top of a roller coaster's tallest hill, right before it plunged him downward for the rest of the ride. That's what life was doing; he was in the split second before it plunged him onward.

He took comfort in the sight of the other Humvees and AAVs in front and behind him. He was part of the world's best-trained and best-equipped fighting force and things would be okay.

Before he knew it, there were buildings to his left. His attention wasn't fully on them. The fact that he was actually riding into battle didn't seem quite real yet. That changed when a bullet fired from one of the buildings ricocheted off the barrel of his weapon. That one bullet was instantly followed by many more.

"Shit!" Inglehart yelled. "We're taking fire! We're taking fire!"

"Shoot back, shithead!" a voice from inside the Humvee said, even though he was already turning the fifty caliber towards the buildings.

The second his weapon was facing the buildings, Inglehart pulled the trigger, unleashing a torrent of angry little projectiles. He couldn't tell which hits were his because the fifties on top of each of the American vehicles were returning fire as well, digging ugly little pock marks into the two-story mud brick and cinder block buildings.

Another bullet ricocheted off the top of the Humvee, nearly grazing Inglehart.

"Come on, man! Drive like this is the Daytona 500 and get us out of here!" he barked at the driver.

No reply came.

Another bullet ricocheted off the side of the Humvee, nowhere near Inglehart this time. The scene in the movie *Black Hawk Down* shot through his mind, where one of the soldiers was manning the fifty on top of a Humvee when he was shot in the head and killed instantly.

"Not me! Not me! Not me!" he yelled with a fury that had never been in his voice before.

Recruit training had broken down the civilian in him and turned him into a Marine. This initial firefight, only the first of many he would be in that day, broke down the peacetime Marine that Inglehart had been and turned him into a true warrior. He didn't think twice about whom the people who were shooting at him were, or if they had wives, children or parents waiting for them to return. They were trying to kill him, which meant only one of them would survive the day and Inglehart was one hundred percent committed to that being him, not the Iraqis firing at him. While he wasn't quite yet battle hardened, the role of true warrior was one that Gary Inglehart slipped into with the ease of putting on an old and well-worn pair of slippers.

Even though it had seemed much longer, Bravo Company was past the buildings in a matter of minutes and the gunfire fell silent. Inglehart's heart was beating a mile a minute and adrenaline coursed through his body in a way he had considered impossible before what he had just experienced. He was alive, more alive than he had ever been in his life.

Staff Sergeant Dave Ligget sat in the rear seat on the driver's side of the Humvee. Unlike most of—all of, as far as he knew—the Marines in Bravo Company, this was not his first time in combat. He had already completed a tour of duty in Afghanistan before being transferred to his current assignment.

Unlike Inglehart above him, Ligget had had his M-16 pointed out the window and at the buildings before they had come upon them. He had been through this before and knew to expect danger from anywhere at any time. He fired short and controlled bursts at the open doors and windows of the buildings as they drove pass them.

The brief firefight might have energized Inglehart, but it had pissed Ligget off. He hadn't liked being shot at while he was in Afghanistan and being in Iraq hadn't done anything to make him like it any better.

"Did you see that?" he heard Inglehart say excitedly. "Did you see that? There were bullets bouncing off of the Hummer all around me."

"It's not done yet!" Ligget barked. "Keep that fifty ready and keep it pointed at anyplace Haji might be hiding!"

A second passed.

"Right, Sarge. Sorry," Inglehart replied. He sounded humbled.

Ligget wondered briefly if any Marines had been hurt or killed. There wasn't any way for him to know and he didn't have long to think about it because they were on the bridge quickly after passing the buildings.

No sooner had they reached the bridge when the Iraqi fire started back up again. Ligget saw it was coming from a pontoon footbridge 1,000 to 1,500 yards away from the one they were driving over.

Above him, Inglehart opened up with the fifty-caliber as Ligget fired controlled burst at Iraqis on the footbridge. He was oblivious to the burning hot spent shell casings that rained down on him from the fifty-caliber above.

Once again, Ligget and Inglehart responded differently to being under fire. While Inglehart was yelling taunts at the Iraqi soldiers, Ligget was grimly silent and coldly professional.
Back at the buildings, the Iraqis had only fired AK-47s at them, but on the bridge he also saw the smoke trails from rocket propelled grenades (RPGs) coming at them as well. "We've got RPGs!" Ligget warned the driver, who didn't say anything in reply.
The RPGs were fired wildly and not a single one even came close to hitting the Humvee he was in.

Damn, he thought. *I hope every Haji shoots this badly.*

He saw several Iraqis on the footbridge fall and not get back up. There wasn't any way for Ligget to know if he had killed anyone or not but he was certain he had and the thought didn't bother him in the least.

It only took Bravo Company a few seconds to cross the bridge. Where there had been a brief respite between the buildings and the bridge, they weren't granted any such luxury once they were off of the bridge and in the city proper.

"Welcome to El Jasiph!" Inglehart heard someone yell from inside the Humvee, but he couldn't tell who it was. He saw muzzle flashes coming from the windows and roof of a mosque ahead of them. Since he didn't see any more immediate threats to his well-being, he turned his weapon towards the mosque and started firing at its roof.

CHAPTER SIX

Alpha Company Privates First Class Leonard Karn and Chris Martinez sat in the track they'd been assigned to, unable to see what was going on outside. They were hot, sweaty and, after being cooped up in the vehicle for so many days, cranky and irritable, but Martinez did his best to be friendly with the Marines around him. The Marines hated riding in tracks, especially while they were traveling across a hot and dusty desert.

None of them had had much sleep since leaving Camp Nicholas and Karn dozed lightly, doing his best to get in as much rest as he could. Given the heat inside the track and the conversations taking place around him, he wasn't having much luck getting any substantial amount of sleep, or even a good power nap.

Lance Corporal Barry Queen watched the miles of desert pass by them through one of the track's firing hatches. He didn't like riding in the track any more than anyone else, but he was grateful to at least have a view of the outside. Being a lifelong fan of the Star Wars movies, Queen passed the time by imagining he was Luke Skywalker being taken across the sands of Tatooine on one of Jabba the Hutt's skiffs to the Sarlacc Pit for execution in *Return of the Jedi*.

The first pinging caused by Iraqi bullets bouncing off the exterior of the track caused all conversation to stop and snapped those who had been daydreaming back to full alertness. Queen, along with the other Marines manning the track's firing hatches, brought their weapons around and began returning fire.

The sound of weapons being fired snapped Karn out of his doze. "What's that?" he asked with just a touch of panic in his voice.

"Nothing for you to be concerned about," Queen answered. "It's just Haji saying, 'Welcome to the neighborhood,' and me replying, 'Pleasure to be here you camel humpin' sons of bitches.'"

His reply caused an uneasy laughter among the men around him, especially from those who weren't at a firing hatch and felt helpless to do anything other than wait and see what the fickle whims of fate had in store for them.

Queen saw a geyser of sand erupt, well away from them, as an Iraqi mortar hit the ground.

"Mortars! The bastards are firing mortars at us!" he heard someone else announce.

"Where are they? I don't see them!" a Marine manning a firing hatch on the opposite side of the track from Queen said.

"My side!" Queen answered, as he fired his M-16. "They're firing from the top of oil tanks in the refinery on my side!"

Marines in Alpha Company's command track also returned fire. The Company's Commander, Captain Dave Callen, moved from hatch to hatch, doing his best to take stock of the situation that his company now found themselves in.

He was just as scared as the Marines he led; like most of them this was his first time under fire. Fortunately, Callen had been well trained and appeared to be every bit the cool, calm and collected Marine Corps officer that everyone expected him to be.

"They anywhere else besides the refinery?" he asked.

"No, sir, doesn't appear so," one of his Marines answered.

Callen looked out the hatch on the refinery side of the track. His jaw dropped when he saw sand geyser after sand geyser erupting from the ground between his company and the oil refinery.

Looks like Hell's answer to Yellow Stone Park, he thought as he saw the mortars striking the ground and blowing up.

"They're walking them up on us!" Callen yelled, more to himself than to anyone in particular.

Mortars aren't like missiles; they can't just be aimed, locked on to a target and fired. One had to be fired and then the targeting information had to be adjusted to get the second shot closer to the target. Then the person firing the mortar had to readjust his targeting information so that the third shot would land closer than the first or second shots had, and each shot went like this until you hit your target. This is called walking.

Why isn't the CAAT hitting them with their TOW? he wondered.

CAAT stood for Combined Anti-Armor Team and consisted of four Humvees. Two of the Humvees were armed with fifty-caliber machine guns, one with a Mark 19 grenade launcher that fired grenades as rapidly as a machine gun fired bullets, and one with a TOW weapons system, a missile launcher.

As if the machine gun fire wasn't bad enough, the mortars had the potential to decimate his company. They were a deadly rain that Callen did not want falling on his convoy of tanks, tracks and Humvees.

He looked at the refinery again to figure out exactly where the mortar positions that were targeting his men were and saw they were coming from three large oil tanks strung out the length of the refinery.
M-16s and fifties aren't going to be enough to take out those mortar positions. We've got to take out those oil tanks or a good share of my Marines won't make it long enough to set foot in El Jasiph, Callen thought.

Alpha Company's Commanding Officer walked to the front of the track, picked up the mic to the radio and checked to make sure it was set to the Company TAC, which would allow him to speak to every vehicle in his convoy. His initial thought was to order the CAAT's TOW operator to take out the oil tanks. Callen decided against that because he didn't know how bad things were going to be in El Jasiph and if they did get bad, he would need those TOW missiles in the city.

Instead of ordering his TOW operators to fire on the oil tanks, Callen switched the radio to the battalion tactical operations command, or TAC, the radio network that would allow him to speak with anyone involved in the day's battle and not just his own company. Callen thought it would be best to save his CAAT's TOW missiles for when they were truly needed, so he decided to call the FiST track.

FiST stood for Fire Support Team and were the forward observers. If a platoon or company needed heavy bombing somewhere, they called the FiST track, which would send help in the form of air cover, both fixed wing (jets or bombers) or rotor (helicopters), and eighty-one millimeter mortars.

"Las Vegas, this is Bigfoot," Callen said into the radio's mic.

"Go Bigfoot," the forward observer's voice replied.

"Roger, requesting rotor wing support."

"Roger, stand by," the forward observer said.

A few seconds of silence, which seemed to be much longer to the Marines of Alpha Company, passed, then the forward observer on the FiST track asked, "Bigfoot, where do you need the support?"

"At the oil refinery on the east side of the city before the bridge. We have Iraqi mortar positions on top of all three oil tanks," Callen answered.

"Roger Bigfoot, rotor winged air support is inbound," the forward observer on the FiST track replied.

Those words filled Dave Callen with more relief than he had honestly expected them to. After all the training he had been through, the thirty-three year old captain left the United States and came to Iraq believing he was ready to handle anything and everything that came his way. Now, as scared as he felt with the Iraqis walking the mortars up to his company, Callen wasn't so sure that he was up to the task. Regardless, he knew he had to pretend that he had all the self-confidence in the world. If he couldn't convince himself of it, he had to at least convince his Marines of it.

Even though the Iraqis were still walking the mortars up to his company, just the knowledge that air support was inbound made him feel a whole lot better than he had just a few minutes before.

Callen walked back to where some of his men were still exchanging fire with the Iraqis hidden up on top of the oil tanks. Once back there, he bent over and looked out the side hatch that faced the oil refinery. He had learned what it was like to get shot at and now he was looking forward to learning what it was like to see his air support kick ass up close and personal.

Lance Corporal Queen ejected the empty magazine from his M-16, pulled a fully loaded magazine from his gear, tapped it on his helmet to make sure all the rounds were seated properly and inserted it into his weapon. With his weapon freshly loaded, Queen pointed it back out the firing hatch and resumed trading bullets with the Iraqis in the oil refinery.

What the hell? Queen thought when he heard the *whump, whump, whump* sound of incoming helicopters.

As soon as he realized what the sound was, one thought went through his mind. *Oh dear God, please let those be ours.*

A couple minutes later he saw two AH1Z Super Cobra helicopters coming in fast and low, aiming themselves right at the oil refinery. He stopped firing.

"Hell fuckin' yeah," Queen muttered at the sight of the rapidly approaching attack helicopters. The mortars had started falling too close for his comfort and, at that point in time, he couldn't think of any sight more beautiful than the incoming helicopters.

"Hell fuckin' yeah!" he yelled. "We've got Cobras!" He turned around to look at the Marines inside the track. "Did you hear me? We've got Cobras!" he said, excitement filling his voice in a way that it never had before.

Queen turned back around. He wouldn't have let anything cause him to miss witnessing the Cobras taking out the Iraqis on top of the oil tanks.   The Cobras came in from different sides. They each fired a Hellfire missile almost simultaneously. The Marines who witnessed it thought the attack looked almost choreographed and well-rehearsed.

Their missiles came in hot and fast, hitting their targets exactly as the pilots who had fired them had intended. Both oil tanks went up in a quick burst of flame, which became a steady fire as the oil contained within ignited and burned. As it burned, the oil released a thick, black smoke that drifted upwards into the sky, the smoke signal let the Iraqis waiting for them in El Jasiph know that the U.S. Marines weren't going to roll over and play dead for them.

Burn in Hell Haji, Queen thought as he saw the two oil tanks go up in flames.

Even though the incoming Iraqi fire from the top of the last remaining oil tank had fallen silent, one of the Cobras circled back around, locked a missile on it and fired. The missile raced away from the helicopter and towards the oil tank, striking and destroying it with an explosion that wasn't any less spectacular than the two before it had been.

A cheer of joy went up in the track. Someone yelled, "Score one for Uncle Sam's Misguided Children!" The nervousness that many of them felt just a few minutes before was replaced by excitement. They were now eager to face whatever waited for them in El Jasiph.

Both helicopters circled around again and came in low and fast against other targets in the oil refinery, opening up on them with their Hellfire missiles, their rocket pods and their twenty-millimeter, three-barreled Gatling guns.

"What the hell are they doing?" one of the Marines asked.

Queen shook his head, mystified. "They must have seen more of something in there waiting for us. Thank God they did."

The Cobras made a few more attack runs before returning to base. By the time they pulled out, the entire oil refinery lay in a fiery, smoky ruin and the Marines of Alpha Company were energized and ready for more.

CHAPTER SEVEN

Staff Sergeant Paul Gates felt a lump of nervousness develop in his stomach as they approached the bridge they would have to cross to enter El Jasiph.

Things had not been going great for him personally. He had enlisted in the Marines several years before, choosing the Marine Corps over the Army mostly just to annoy his brother, Stanley, who was sixteen years older than he was, and who had been a career soldier. Stanley had been killed in action while serving in Afghanistan.

After Stanley's death, his parents had been after Paul to quit the Marines. It wasn't as if he could just walk up to Captain Aber, his Company Commander, and just say, "I quit." Paul had tried to convince his parents that it wasn't that easy; he didn't work for McDonalds. But, in their parental worry, they didn't listen to him and kept insisting that he quit.

As if Stanley's death and the family stress that he was under weren't bad enough, three months after arriving to Camp Nicholas, Paul's wife had sent him a letter telling him she was filing for divorce. She had said, "It's not you, it's me. I just can't handle you being so far away from me. I need a man who is going to be able to be home and sleep in bed with me every night and you've just failed to be there." Well, what did she expect? He had been a Marine when they met. She enjoyed going to the annual ball that marked the Marine Corps' birthday with him and when he had been a Drill Instructor at the Marine Corps Recruit Depot San Diego, she had enjoyed showing off her husband, "the Drill Sergeant." What had she expected, that his job wasn't anything more than yelling at young recruits and looking handsome in his uniform? Marines were trained for war. He had thought she understood that.

So Gates had a lot more on his mind than just what might happen to him in El Jasiph, or what might happen to him during the rest of his time in Iraq. Most of all, Paul just wanted to conduct himself in a way that would make Stanley proud of him. He knew his older brother was up in Heaven watching him and the last thing he wanted to do was disappoint his fallen brother.

He could hear weapons firing in the distance as Alpha and Bravo came under attack. The sound wasn't one he liked, but it did help steel him for whatever was to come. It also caused him to scrutinize the houses of the outlying village that Charlie Company was passing through all that much closer.

However, he didn't see anyone coming after them with AK-47s or RPGs. Instead, he saw average, every day Iraqis going about their daily business. Instead of having mortars or artillery shells falling on them, Paul Gates saw Iraqis waving at Charlie Company as they drove through and waving white flags.

When Stanley was here during Desert Storm, he said the Iraqi Army gave up quickly.

Maybe things won't be so bad for us either, Gates thought as he watched the Iraqi civilians around him. The thought caused the nervous lump in his guts to shrink quite a bit, but it refused to go away completely.

Gates looked away from the firing hatch and towards the other Marines in the track. "They're waving white flags," he said.

"This is going to be a piece of cake," a corporal, named Chris Jabcon, said. "There might be a few little skirmishes, but Baghdad is going to fall without any real fighting. Haji doesn't like Saddam all that much."

He looked back out the firing hatch and looked at the smiling faces of the Iraqis. *They don't seem scared of us at all. In fact, they seem happy to see us. Yeah, maybe this won't be too bad after all.*

CHAPTER EIGHT

Passing the mosque didn't mean Bravo Company would get the brief respite they had gotten after passing the small collection of buildings on the other side of the bridge. They didn't, and they would be in a running firefight for the next several hours, which to the Marines of Bravo Company would seem more like days.

The city's streets were lined with two-story houses made of mud brick and cinder block. Some of the houses had courtyards and others didn't. Walls that were eight to ten feet tall surrounded the courtyards. This set-up gave the Iraqi soldiers the best cover they could have asked for while they fired at the U.S. Marines.

The convoy was led by three tanks, and the CAAT Humvee with the TOW weapons system and one of the CAAT's fifty caliber Humvees. The rear was brought up by the CAAT's other fifty caliber Humvee and the Humvee with the Mark 19 grenade launcher. The tanks and the CAAT Humvees, along with the four snipers who were sitting in one of the tracks, were on loan to Bravo Company from the Battalion's Weapons Company and weren't part of the company's regular compliment of Marines.

Originally the plan called for Bravo Company to make the first right once they were over the bridge but the lead tank missed that turn and took the second right instead. Shortly after crossing the bridge, the already bumpy ride in the tracks became just that much worse for the Marines penned up inside of them.

Sergeant Duane Owen fired his M-16 out the back window of the Humvee he rode in. His eyes grew wide when he saw the smoke trail of an RPG come streaming out of a second floor window of a house. The RPG flew over his Humvee, too high to have been a danger to any of the vehicles in the convoy. He heard it hit one of the buildings on the other side of the road, but he was too busy firing his weapon to take a look.

"Holy shit! Did you see that?" Lance Corporal Glenn Utter, who was sitting next to Sergeant Owen, asked.

Owen didn't reply. He knew the question had been asked out of surprise and wasn't an attempt at conversation. He saw an Iraqi, wearing civilian cloths, but pointing an AK-47 at the convoy; step out from behind a courtyard wall that looked like one of the tanks had already fired at it.

Cowards are dressing like civilians; Owen thought when he saw the Iraqi soldier.   He aimed his M-16 at the Iraqi and squeezed off a controlled burst. His bullets struck the Iraqi in the chest and he toppled over backwards. Owen had fired quite a few rounds since Bravo Company entered El Jasiph, but it had all been on the drive fighting and that Iraqi soldier was the first one he could say for certain he had hit.   That was when several bullets struck the Humvee. Most of them hit the vehicle's doors and ricocheted harmlessly away from the Marines inside. One of them shot through the window that Sergeant Owen was firing out of, crossed the Humvee and exited through the window that Lance Corporal Utter was firing out of. The bullet had come so close to hitting Owen that he had actually felt the air movement caused by its passage on his right cheek.

Stunned by the close call, Owen quit firing for a second and looked back at Utter, who was already looking at him. The expression on Utter's face mirrored his own dumbfounded expression exactly.

"This shit's real!" Utter said.

"Goddamn right it is," Owen replied, already turning back around to continue fighting. "As much as I hate riding in 'em, I'd give my left nut to be in one of the tracks right now. More protection than this Hummer," Owen said.

"Too thin walled, not designed for this kind of fighting," Utter replied without looking at him.

"The hell with you." Those guys got more protection than we do right now," Owen said. "What I'd really like to be in right now is one of them CAAT Hummers," The standard company Humvee, like the one they were in, had cloth coverings that offered some protection from the elements to those inside of it. The CAAT Humvees however was a hardback, meaning that its top was steel and not cloth, to help protect its occupants from more than just the elements. They also had sandbags on the floor as an added line of protection from enemy bullets that the standard Humvee didn't have. While the Marines in the standard Humvee were left to the fickle whims of fate, steel and sand surrounded the guys assigned to the CAAT.

From above them, Private Larry Rolling, who was manning the Humvee's fifty caliber, said, "Fuck both ya'll. Ya'll got more around you than I do up here."

Iraqi bullets struck the top of the Humvee all around Rolling. He had also felt the air movement caused by bullets coming way too close for comfort many times over.

Rolling saw an Iraqi on top of one of the two-story houses putting an RPG launcher up to his shoulder.

"Not today you don't Haji," he muttered without even realizing he had.

He aimed his fifty-caliber at the Iraqi with the RPG launcher and pulled the trigger. The Iraqi's head disappeared in a gore filled, pink cloud and the rest of his body dropped onto the house's roof. He didn't get back up.

Even though he was in combat and his life was in the most danger it had ever been in during his eighteen years of life, something about the whole experience didn't seem quite real to him. It felt almost as if he was in the most realistic, virtual reality first-person-shooter game ever created.

His conscious mind knew that he should have been scared, that he should have been wetting his pants, but his blood was up. The only thing that kept him from feeling fear was the adrenaline that was coursing through his veins. Rationally, Rolling knew that he shouldn't be, but there was a part of his brain that refused to admit that the situation was truly real and was actually enjoying the experience.

Inside one of Bravo Company's tracks, Corporal Dan Abraham stopped firing his M-16 out the track's side firing hatch just long enough to wipe the sweat away from his face. Like most American males, he had watched the Rambo movies repeatedly and he couldn't ever remember seeing Sylvester Stallone sweat as much as he was, but then, he had also never seen Sylvester Stallone fighting in a suit designed to protect him from nuclear, chemical and biological weapons either.

With the sweat off his face, he placed his weapon's barrel back out the hatch and started looking for somebody to shoot. Abraham saw an Iraqi, this one wearing his Iraqi Army uniform and not masquerading as a civilian, pop up in a window and fire his AK-47. Abraham returned fire but by the time his bullets hit the window, the Iraqi had already ducked out of the way.

How am I supposed to hit people who won't stand and fight? he thought, while at the same time wishing he could be using the same hit and run tactics the Iraqi soldiers were using.

Another Iraqi wearing civilian clothing stepped out into the open. He was of military age but Abraham didn't see a weapon so he held his fire. The fact that so many of the Iraqi soldiers were fighting in civilian clothing made it hard not to shoot at everyone he saw. "Hold still, will you?" Abraham heard someone say.

For some reason, even the bullets flying through the air couldn't keep him from smiling. *At least I'm not the only one having trouble with them,* he thought. *It'd be nice if God would rain down super glue or something on top of them.*

"That would take all of the fun out of it," another replied.

"Just like *Duck Hunt* on the Nintendo I had as a kid," someone else said.

The comment caused a couple of uneasy chuckles. The attempt at humor may have seemed out of place considering there were thousands of Iraqi soldiers trying to kill them, but it helped to lighten the tension.

Abraham emptied his magazine and ejected it from his weapon. He pulled out another one, tapped it against his Kevlar helmet and inserted it into his M-16.

No sooner had he pointed the barrel of his rifle out the firing hatch, that Abraham saw an Iraqi, in civilian clothing, standing in a window, firing his AK-47 without bothering to hide in between shots.

He sighted in on the Iraqi and fired off a short burst. The Iraqi grabbed his stomach as if it were cramping really badly, doubled over and disappeared from view.

"Just got me a duck!" Abraham announced.

His announcement brought a round of cheers from the other Marines. Especially from the ones who didn't have a hatch to fire out of and were just riding along, hoping the enemy bullets didn't puncture through the track's thin hull and hit them.

At the rear of Bravo Company's convoy, Sergeant Scott Dove manned the CAAT's Mark 19 grenade launcher.

Shit, he thought. *In the movies they always show the Marines on one side and the bad guys on the other shooting at each other. This is one of the biggest cluster fucks I've ever seen. I can't tell where most of the shooting is coming from.*

A streak of black smoke told him that an RPG had been fired. Like all the others that had been fired at the convoy so far, the RPG was poorly aimed and missed.

Thank God these guys can't shoot. They might be dangerous if they could, Dove thought as he turned his head towards the area where the RPG had been fired. He saw an Iraqi with an RPG launcher on his shoulder, ready to fire another one.

"No you don't," Dove mumbled as he turned his Mark 19 and sighted in on the rooftop where the Iraqi stood.

He pulled his trigger and sent eight rapidly fired grenades flying towards the roof. All eight landed on target and all eight exploded. It filled Sergeant Dove with a grim satisfaction when he saw the armless upper body of the Iraqi who had been firing the RPG fall from the roof and land on the unpaved street.

"Enjoy your virgins motherfucker," he grumbled as he looked for another target. Muzzle flashes coming from an upstairs window gave it to him.

Dove sighted in the window and pulled his weapon's trigger. This time, six grenades flew through the air and all six of them entered the window.

He was rewarded by seeing six distinct flashes as each grenade exploded, shrapnel cutting anyone in the room to ribbons. Smoke wafted from the room.

CHAPTER NINE

Alpha Company came under heavy attack on the El Jasiph side of the eastern bridge. The Iraqi bullets that hit and ricocheted off their vehicles weren't the real cause for concern because no one had been hit yet, the RPGs being fired at them were. Just one of them would have been more than enough to kill several Marines. The RPG fire was so heavy the convoy's drivers were forced to put their vehicles into a zigzag motion to avoid being hit.

An Iraqi soldier, in civilian clothing, stood up in a sandbag bunker with an RPG launcher up to his shoulder, ready to fire.

Lance Corporal Victor Lauralwood manned the Alpha Company CAAT's Mark 19 grenade launcher and saw the Iraqi behind the sandbags. With bullets striking the top of the Humvee all around him, Lauralwood turned his Mark 19 and fired off five rapid-fire grenades before the Iraqi could fire his RPG.

Each of Lauralwood's grenades hit home and the bunker blew up in a fog of sand. Once the fog cleared, Lauralwood saw that all that remained of the Iraqi with the RPG launcher was a gory, mangled shell of a man. The pieces of the corpse, which lay scattered around what had been the bunker, no longer resembled any piece of a human body.

He saw another RPG streak towards the convoy. This one barely missed striking the side of one of the AAVs ahead of him. He looked towards where the RPG had been fired from and saw a uniformed Iraqi soldier loading another grenade into his launcher in a first floor window.

Lauralwood swung his Mark 19 around and aimed the weapon at the window. He pulled the trigger and several more grenades fired. Not all of them went through the window—some struck the side of the mud brick house and blew up, knocking large holes in the wall as they did—but most of them hit their mark. Smoke drifted out of the window and, even though Lauralwood didn't see the body he was confident that he had eliminated that particular threat.

Son of a bitch, Lauralwood thought as he saw another trail of black smoke from a RPG. *There's so many of those fuckers I can't even think about taking out the guys with guns.*

In the very last of the convoy's Humvees, right behind Victor Lauralwood, Private First Class Brad Barnebee manned one of the CAAT's fifty caliber machine guns. At twenty years of age, his mind was panicked and he fired his weapon at pretty much anything that moved, in buildings, on rooftops, on the street, anywhere, holding his fire only for women and children.

Iraqi bullets were hitting his Humvee from everywhere and, in his less than optimal state of mind, he was having trouble telling where the enemy fire was coming from.

I should have gone to college like my momma wanted me too, Barnebee thought as he stitched bullets along the cinder block courtyard wall of a two-story house. *This was supposed to be exciting. This just sucks!*

A battered pickup truck with a Vietnam-Era M-60 mounted on a tripod in its bed turned a corner, out of an alley, and took up position right behind Barnebee's Humvee. Besides the M-60 gunner, there were two other men in the truck bed who were armed with AK-47s. The only Iraqi in the truck's cab was its driver.

All three Iraqi gunners started firing on the American convoy the second they were behind it. The M-60's gunner was not very skilled and his bullets kicked up dirt behind Barnebee's Humvee, but the young private wasn't naive enough to believe the man's aim wouldn't improve quickly.

Barnebee turned his fifty-caliber towards the pickup truck and fired at the M-60 gunner, striking him in the chest. The man's torso seemed to disintegrate in a bloody cloud. His lower body collapsed in the truck bed while his arms and head fell clear of the truck and came to rest on the unpaved road.

One of the Iraqis with an AK-47 dropped his weapon and moved to take over the M-60. *No you don't,* Barnebee thought as he fired again.

This time his bullets didn't hit the Iraqi, but they did get his attention and caused him to dive out of the pickup truck, like a swimmer entering a swimming pool, to avoid being shot. Barnebee didn't see if his escape had killed the Iraqi or not, but figured if he still lived, he had to be severely hurt after a fall like that.

You guys just don't learn, Barnebee thought as he saw the truck bed's last remaining occupant drop his AK-47 and move for the M-60.

He fired his fifty-caliber again, only this time instead of aiming at the man operating the M-60, Barnebee aimed at the truck's driver. The driver disappeared in a sudden explosion of red and the truck veered sharply to the left. The sudden motion threw the Iraqi in the truck bed clear of the vehicle and he smashed into a mud brick house face first, breaking his neck and killing him instantly. The out of control pickup truck punched through the wall and came to a stop somewhere inside the building. The last Barnebee saw of the truck were the flames from the fire the crash caused.

As his Humvee drove past an alley, Barnebee saw another M-60-equipped pickup truck coming towards the American convoy as fast as it could. This time, Barnebee just fired his weapon at the cab, punching out the windshield. He would never know what happened to those particular Iraqis, but Barnebee did see the speeding pickup truck go out of control and start to roll before his Humvee passed the alley.

An ambulance, with its lights and sirens going, sped towards the tanks that led Alpha Company's convoy. It came to a stop and several men in civilian clothing poured out of it, all of them carrying AK-47s.

The M1A1 Abram's cannon fired a one hundred and twenty millimeter shell, striking the ambulance. The Iraqi vehicle exploded, killing and injuring several of its former occupants with the large pieces of shrapnel that flew through the air around it.   Most of those who had managed to avoid being killed or severely wounded by the flying shards of metal decided they had had enough and ran to the shelter provided by the surrounding buildings. Some of them carried their AK-47 with them, others left their weapons lying on the street as they fled. There were a few Iraqi soldiers who stood their ground and started firing at the lead tank, as if they thought they had chance of pulling a David and Goliath-type of upset by taking out an American tank with their AK-47s. Those who didn't run were reduced to bloody chunks of meat by the fifty-caliber machine gun mounted on top of the tank

Lieutenant Colonel Everett, the Battalion Commander, had given up smoking a cigar when the first one was shot out of his mouth by a passing Iraqi bullet.

This is bad, Everett thought, taking stock of the situation that his Marines found themselves in, *and has the potential to get a lot worse.*

He had been listening to the Battalion TAC and knew Bravo company was under heavy fire and Charlie Company wasn't yet, but assumed they probably would be as soon as they crossed the northern bridge. He knew Alpha Company was under heavy fire because he was under heavy fire with them.

These Iraqis can't hit the broadside of a barn right now, but sooner or later their aim is bound to improve, or they'll just get lucky and they'll start scoring hits. This ambush is too heavy for us to keep pushing through without starting to loose people. Our orders are to make El Jasiph safe for those who will come through after us and we can't do that without getting out of the vehicles.

Everett thought about the situation a little longer, while hearing Iraqi bullets ricochet off his Humvee and hearing the fifty-caliber that Corporal Bruce Hawkins was manning fire almost nonstop.

He turned around to look at Sergeant Major Art Jamison in the backseat. "Sergeant Major, get Captain Callen on the horn and tell him to bring the company to a stop. Trying to drive through this will only get us torn up before we can get into position to support Charlie," Everett ordered, yelling to be heard over the noise of the battle that surrounded them.

"Yes, sir," Jamison replied and turned his attention from firing his M-16 to operating the radio.

The AAV Commander stood and faced the rear of his vehicle. Privates First Class Chris Martinez and Leonard Karn, Lance Corporal Barry Queen and the nine other Marines in the track all looked at him with expectant eyes.

"Ok, boys, get ready, we're going to be stopping and getting out," the track commander said.

All throughout Alpha Company's convoy, Marines were checking their weapons over real quickly and preparing themselves mentally for whatever lay ahead of them.

The AAV came to a stop and its commander hit a button. The rear ramp began to lower and the Marines inside stood. They were all nervous, but all the training and drilling that they'd been through took hold. As soon as the ramp lowered fully, they charged out, their weapons up to their shoulders. Some of them laid flat on the ground and others found cover where they could. All of them, now fully exposed to the enemy, began returning fire.

CHAPTER TEN

Captain Brain Aber, Charlie Company's Commanding Officer, looked out the front of his command track as it came over the bridge and entered El Jasiph. The fact that Charlie Company had not come under attack prior to crossing the bridge had made him overconfident that the Battle of El Jasiph would be short and sweet.

Sure, he heard the sounds of weapons being fired in other areas of the city, but thought of it as nothing more than small pockets of Saddam loyalists in the Iraqi Army making their last stand instead of either deserting or surrendering. Aber thought it was likely the Fedayeen, the paramilitary arm of Saddam Hussein's Ba'athist Party, which meant that, unlike the other Iraqi soldiers, their loyalty was to the Ba'athist Party and Saddam Hussein, not to the country of Iraq. The Fedayeen would be just as likely to kill an Iraqi as an American if they tried to surrender or showed any disloyalty to Saddam or the Ba'athist Party.

Alpha and Bravo Companies had both come under fire before crossing their bridges into El Jasiph and while Alpha had been able to cross their bridge unmolested, Bravo had had to fight their way across their assigned bridge. Charlie Company, on the other hand, had not been fired upon once.

And they told me I'd have a harder time with this advance than the other companies, Aber thought with a silent chuckle. *My company is the only one that isn't being shot at.*

He watched the drab, almost identical looking mud brick and cinder block houses as they drove deeper and deeper into the city of El Jasiph with a growing sense of excitement.

Well, we're not in America any more Toto. This place is so backwards that it almost feels like we've been sucked back into time, he thought as he looked around at the primitive, by American standards, looking city.

The last CAAT Humvee, which brought up the rear of Charlie Company's convoy, drove off of the bridge. Sergeant Rick Neighbors manned the last Humvee's fifty-caliber, and he didn't share Captain Aber's optimistic outlook on how well the day would go. He wasn't being pessimistic and assuming the entire battalion would be wiped out here either; he knew that would be the least likely outcome of the day. They were simply too well armed and too well trained for the Iraqi Army to wipe them out. But that didn't mean that they couldn't hurt them. That would be an incredibly arrogant assumption, so Neighbors just adopted a let's-wait-and-see attitude.

Less than a minute after his Humvee pulled off of the bridge, Neighbors saw an Iraqi in civilian clothing standing at a corner and one word came to his mind to describe the man's actions, lurking. The Iraqi kind of reminded him of those old commercials that used to air during Saturday morning cartoons, the ones with the strange guy hanging out around a school in a trench coat that warned children of the dangers of strangers. Neighbors kept an eye on this particular Iraqi; he didn't trust him.

A couple seconds later Neighbors saw the Iraqi move and expose the AK-47 he was carrying. He pulled the weapon up to his shoulder and pointed it at a vehicle in the convoy well ahead of the Humvee Neighbors was in. Neighbors already had his weapon pointing at the Iraqi and fired before the enemy soldier could pull his trigger. The bullets struck the Iraqi just above his waist and tore the man into two halves.

Neighbors pulled his trigger at the same time the Iraqis sprang their ambush. Charlie Company's convoy was under attack from everywhere.

An Iraqi pickup truck, with an M-60 machine gun mounted in the bed pulled out of an alley and laid down a heavy steam of fire at the convoy. Unlike the other modified pickup trucks the Marines had seen that day, this one only had one driver in the cab and one soldier in the back of the truck. Neighbors aimed his fifty-caliber machine gun at the Iraqi M-60 operator and fired.

His bullets raced through the air to their target. Some of them struck the enemy soldier's weapon and ricocheted off. None of his shots scored a kill but he did hit the Iraqi in his right wrist, separating the man's hand from the rest of his body.

Shock at losing his hand caused the Iraqi soldier to back up, which threw him off balance and he fell out of the bed of the pickup truck, coming to a stop as a crumpled wreck of a man on the unpaved street.

The pickup truck's driver realized he had lost his gunner and instantly knew he was in trouble. He slammed his foot on the break and the truck slid to a stop. After coming to a stop, he turned the pickup around and started running away as fast as his battered and worn out truck would drive.

Neighbors fired again, striking the rear of the truck several times, including a direct hit on the gas tank. The pickup truck disappeared in a ball of flames and rolled to a stop.

In one of the company's tracks, Sergeant Ed Paige fired his M-16 out the track's rear firing hatch. Paige was known as Sergeant Fuckin' throughout the entire battalion because of the fact that he seemed to be incapable of completing an entire sentence without using the word fuckin' at least once.

When his first magazine ran empty, he pulled it out, tapped the new one on his Kevlar helmet and loaded the fresh magazine, which was loaded with bullets.

"This is like fighting in a fuckin' maze!" Paige yelled.

"What was that?" one of the other Marines in the track asked.

"I said, this is like fighting in a fuckin' maze. These motherfuckerscan be hiding any-fuckin'-where," he replied.

On top of one of the buildings, Paige saw a woman hanging wet clothes on a clothes line.

What's that fuckin' bitch doing? Who hangs fuckin' cloths in the middle of a fuckin' firefight? he thought.

Civilians killed Army Rangers during the Black Hawk Down incident, so Paige kept his eye on the woman for a few seconds just to make sure she didn't whip out an RPG launcher or something. After a few seconds, he was certain she wasn't anything more than she appeared to be, an innocent local woman simply hanging her wet laundry out to dry.

Paige then saw an Iraqi in the Iraqi Army uniform step out of a doorway. He aimed his M-16 at the Iraqi and pulled the trigger. His bullets struck the enemy solider right around his diaphragm and stitched their way to just below his throat.

The Iraqi soldier fell over backwards and didn't move again. Paige started looking for his next target.

CHAPTER ELEVEN

Bravo Company pushed through El Jasiph under a continuously heavy onslaught of enemy fire, which they were not stingy in returning. It seemed the Iraqis were throwing pretty much everything in their arsenal at them, machine guns, RPGs and mortars. Apache attack helicopters had taken out most, if not all, of the tanks in the city prior to the Marine infantry's arrival. So far, the Iraqi forces had not used any chemical or biological weapons, as it was feared they might. The Marines engaged in the battle had too much on their minds at that point in time to even think about the NBC suits they were wearing, the same suits that had caused them so much discomfort over the last few days.

Despite the angrily fired bullets flying by them, not a single Marine had received so much as even a glancing flesh wound. The Iraqis couldn't say the same. The difference in training was showing and the home team suffered heavy losses as a result. The problem was there were so many more Iraqi soldiers than United States Marines that they could afford those losses and had plenty more soldiers to throw into the fray.

Directly behind the three tanks and two of the CAAT Humvees which escorted Bravo Company, Captain Dan Earl, the company's commander, listened to what was happening around him on the Company TAC. The situation was much worse than he had hoped, but they were trained to overcome problems and adapt to ever-changing circumstances, and that is exactly what he and his Marines would do.

"Sir, we're coming to a stop," the track's driver said.

Earl looked up and saw they were slowing down. He picked up the radio mic. "Jolly Green Giant, this is Thor," he said, calling the tank in front of the column.

"This is Jolly Green Giant. Go, Thor," the tank commander's voice replied.

"Why are we slowing down?"

"Thor, I'm looking at a wide open area about the size of a city block…" Jolly Green Giant began to say.

"Yes, aerial recon told us there are a few of those in the city. Why are you bringing us to a stop?" Earl asked, cutting him off.

"I'm looking at about a city block's worth of mud. Looks pretty deep."

"Can we make it through it without getting stuck?"

There was a moment of silence.

"I think so. It looks pretty crusted over, I think it'll hold our weight." the tank commander said.

"Then let's get moving forward, Jolly Green Giant. Charlie is going to be in for a world of shit if we don't get over there," Earl said.

"Affirmative, Thor. Jolly Green Giant out."

The column started moving forward at a faster pace again. Captain Earl felt a difference in how the track rode in less than a minute.

God, I hope we don't get bogged down, he thought.

Up until this point, Private First Class Alfred "Big Al" Halik's attention had been focused on driving the Humvee. As he slowed it down to avoid running into the back of the track in front of him, his nose wrinkled at the smell.

"Which of you guys shit your pants?" Staff Sergeant Dave Ligget asked from behind him, mirroring Halik's own thoughts. He asked the question while still firing his M-16 out the window at Iraqis next to him.

"Forget the smell, why are we stopping?" Corporal Pat Nolan, who was sitting next to Ligget and firing out his window, asked.

"Don't know, but it's coming at a real bad time," Lance Corporal Gary Inglehart shouted from above them as he sighted his fifty-caliber down an alley to where four Iraqis had set up a mortar position behind a bunch of sand bags.

He fired and watched, as another Iraqi's head just disappeared in a gore-filled cloud. The enemy soldier had been carrying a mortar shell and Inglehart watched as another went for it. That Iraqi slipped in the pool of blood that was developing around his fallen comrade and hit the ground hard. As annoyed as the soldier may have been, the fall had saved his life because Inglehart had just pulled the trigger again.

All falling had done was bought the Iraqi a few more seconds of life because Inglehart readjusted his weapon and fired again. His bullets hit the Iraqi in the back and his torso disappeared into nothing.

If the convoy had been moving at its original speed, he would have only had the chance to drop the first Iraqi soldier before he would have been past them. However, now that the convoy had slowed down so much, it seemed like he had all the time in the world to take out the sandbagged mortar position.

Commonsense won out over valor and the remaining two Iraqis ran, leaving their mortar tube, shells and fallen comrades where they were. To discourage the Iraqis from returning anytime soon, Inglehart sent several rounds down the alley after them.

Once they were gone, Inglehart began firing at the sandbags, hoping to destroy the mortar tube. If he succeeded or not, he'd never know.

Come on, come on, COME ON! Ligget thought.

At thirty-five years of age, Dave Ligget may have been a staff sergeant but he was a very impatient man and his impatience got the best of him. Seemingly unworried about the bullets flying around, he opened the door of the Humvee and jumped out.

"Staff Sergeant!" Inglehart said from above the Humvee. "Are you out of your mind? Get back in!"

When Ligget didn't reply, or return immediately to the vehicle, Inglehart started providing the best cover fire he could for him.

Hunched low to make a smaller target of himself, Ligget kept moving. The Iraqis didn't fail to notice he was out of the somewhat safe confines of the Humvee and little plumes of dirt kicked up all around him.

Moving in a zigzag motion, Ligget was able to avoid being hit. He only went out far enough to see around the vehicles in front of the Humvee, no more than ten feet or so. *What the...,* he thought, a shocked expression on his face.

Halik saw the track in front of him start to speed up.

"Hey, Sarge, if you don't want to lose your ride you'd better get your ass back here. We're picking up speed," Halik said.

Ligget ran back to the Humvee and got back in as fast as he could. He was met with questioning faces from everyone else in the vehicle. The only one who hadn't quit firing his weapon to look at him was Lance Corporal Inglehart, up top; Halik was too busy driving to look.

"So, what is it?" Nolan asked expectantly.

"You boys aren't gonna believe this," he replied.

"What?" Nolan asked again.

"Looks like the Hajis flooded the clearing ahead of us with sewage," Ligget answered and went back to firing at any Iraqi soldier who made the mistake of showing himself.

Captain Earl felt a knot in the pit of his stomach. He hated the chance he was taking, but he liked his other option, turning the entire company around and fighting their way back through what they had just come through, even less.

Come on, come on, he thought, every bit impatiently as Staff Sergeant Ligget had been before their speed picked back up. Even though the convoy was moving through the urine and fecal matter as fast as it could, it seemed painfully slow. His biggest fear was that his entire company would become mired down in the raw sewage.

He felt his command track start to slow down. Earl moved up to the front and asked,

"Why are we slowing down?"

"This shit's thick and deep, sir. I think we're going to get stuck," the driver answered.

"Well don't let us," Earl ordered.

Do you really think I'd try to get us stuck, dumb ass? the driver thought, but instead said,

"I'll do my best, sir."

A few seconds later, his command track came to a full halt. They were stuck.

The track's driver immediately began rocking the vehicle back and forth, hoping to break it free of the muck.

"Damnit!" Earl yelled as he slammed his fist into the side of the track. As he looked around rapidly, mentally processing their situation and deciding what to do.

"Stop rocking this piece of shit and open up the back. I'm going out to find out exactly what our situation is," he said.

"Yes, sir," the driver replied as he followed the orders.

The putrid odor of raw sewage hit his nose as soon as the rear hatch began to open. Earl realized just how wrong Jolly Green Giant had been in assuming it was mud they were driving through. *Shit? We're stuck in shit? Now this is one that'll follow me throughout my career. This really sucks,* he thought.

The strong odor made him think the mud was two parts water with the rest of it being urine and feces.

Earl walked through the track and looked out the open back hatch. He was happy to see the majority of Bravo Company had come to a stop before entering the sewage. He saw the track following him and the Humvee behind them were just as hopelessly stuck as he was.

Deciding he really didn't have a choice in the matter, he jumped down into the sewage and quickly moved around to the side of his command track. Splashes in the sewage around him made it clear just how heavily the Iraqis were firing on him. He saw that the three tanks and two CAAT Humvees ahead of him were mired down also and couldn't move anywhere. The area they had hoped to cross was the size of a city block; his mired vehicles were spread out over the space about the size of a football field.

"Yeah, this really sucks," Earl mumbled grumpily.

He returned to the rear of the track and climbed back in. As he walked back to the front, the Marines he passed grimaced in disgust at the smell that rose from their Commanding Officer.

"Is it bad, sir?" First Lieutenant Bill Owings asked. Owings was Bravo Company's executive officer, also known as XO, or the second in command of the company.

"Not as bad as it could be," Earl answered. "We've got eight vehicles stuck."

"So, what're we going to do? Leave them?" Owings asked.

Earl shot him a look full of disgust. "Hell no, we're not leaving them for Haji. Get on the radio, get us a tank retriever, tell the FiST to get us some air cover and choose fifty men to stay behind and keep Haji away from our vehicles until we can get them free. I'll have Gunny Schuring send a squad ahead of the rest of us and scout the path, make sure this doesn't happen again. You'll be staying here with the force protecting the vehicles." "Yes, sir," Owings replied.

CHAPTER TWELVE

Forty-year-old Gunnery Sergeant John Ganton was a Marine Corps lifer who considered himself married to the Corps. He had been in serious relationships in the past, but held a deep-seated belief that a U.S. Marine cannot be married and be a Marine at the same time; the two loves divided loyalties and that just didn't work. In all his years of trying, Ganton hadn't found a woman he felt he could love more than he loved being a Marine, so he had more or less sworn off real relationships and contented himself with the company of lower-end call girls.

However, relationships, call girls and even his love for the U.S. Marine Corps were the farthest things from his mind as he stood with a cinder block courtyard wall giving him partial cover. Taking careful aim with his nine millimeter Beretta, Ganton fired at a rooftop where three Iraqis were trying their best to set up a mortar position. The Iraqis were doing more to avoid his bullets than they were to get their mortar tube set up, so even though he wasn't having any luck at hitting them, Gunny Ganton was happy he wasn't making their lives easy up there.

Ganton saw a twenty-year-old lance corporal, whose name he didn't know, curled up into an almost fetal position and leaning against a now mostly empty track.

What's that asshole doing? He's going to get himself tagged if he stays out there in the open like that, Ganton thought.

He watched the young Marine for a few seconds and saw little eruptions of dirt from where Iraqi bullets where hitting the ground all around him. He looked around and saw no one else had even noticed the lance corporal.

Fuck! I'm going to have to go out there and get his ass, Ganton thought as he stopped shooting at the Iraqis setting up the mortar position.

He aimed his pistol at the Iraqi soldiers who were shooting at the lance corporal and started giving them a reason to start being more careful. Once the incoming fire had lessened, Ganton ran, hunched over to make as small of a target of himself as he could, to where the lance corporal was curled up.

"On your feet, Marine! What are you trying to do? Get yourself killed?" Ganton said as soon as he arrived.

The lance corporal didn't even seem to hear him. Ganton shoved his weapon back into its holster, grabbed hold of the young man and hauled him to his feet.

"What's your malfunction, Marine?" Ganton asked, as he noticed the young man was crying.

"I don't... I don't want to die, Gunny. I really don't," the lance corporal answered.

"Well, you're not doing a very good job of keeping yourself alive! If Haji could shoot worth dick, your dumb ass would have a toe tag on it already," Ganton replied as he removed his Beretta from its holster.

The plumes of dirt around them were picking up. Ganton took a second to return fire and turned back to the lance corporal.

"We've got to get back where I was. Grab your weapon, Marine, and let's go," Ganton said.

Suddenly the lance corporal looked embarrassed.

"Gunny?" he asked.

"What? If we don't get our asses out of here soon, both of us will be getting fitted for toe tags!" Ganton replied.

"Uh, I pissed myself."

Ganton looked at him. "I'll bet you're not the first to do it today and I can guaran-damn-tee you're not going to be the last before it's all said and done. I'll cover you while you run to where I was." Ganton pointed the spot out to the lance corporal. "When you get there, you turn around and cover me. Got that?"

"Yes, Gunny," the lance corporal answered.

"Good. Now, move your ass, Marine!" Ganton ordered as he raised his Beretta and started firing at any Iraqi with a weapon.

The lance corporal ran, ducking as low as he could, just as he had been trained, as he did so. When he made it to the spot Ganton had pointed out, he turned around and provided cover fire for the Gunny, which came as a pleasant surprise to Ganton, who hadn't been so sure the lance corporal would have had the presence of mind to follow his order. It was one of the very rare occasions he was happy to have been wrong.

As soon as the lance corporal started providing cover fire for him, Ganton took off running for where he had started out, little plumes of sand following him as he did. As soon as he made it, Ganton looked back to the Iraqis he had been harassing earlier while they tried to set up a mortar position.

Ganton looked just in time to see them drop their first mortar into the tube. It fired, but the shell landed somewhere well behind them. He heard it smash into a building somewhere where fighting hadn't even broke out yet.

He tapped the lance corporal on the shoulder and pointed to the mortar position. "Help me take them out."

"You bet, Gunny," he said. The fear the lance corporal had felt just a few moments earlier was gone and his training had reasserted itself.

They both opened fire on the rooftop Iraqis. There wasn't any way either of them could be sure who shot him, but the Iraqi soldier firing the mortar was hit as he prepared to drop another round into the tube. He fell over backwards and the round landed on the dirt road two stories below where he had stood.

Between the two of them, there wasn't a single living person on top of the rooftop a few seconds later.

Private First Class Chris Martinez crouched next to a car that had been shot up and still contained the corpse of an Iraqi with his hands on an AK-47, exchanging fire with an Iraqi soldier who was firing at him from the second floor window of a house.

I wish I was still in the track, Martinez thought as he pulled the trigger on his M-16 and sent several more bullets flying towards the Iraqi soldier's position. *It sucked in there, but it sucks a lot worse out here.*

Lance Corporal Barry Queen stood behind him firing at Iraqis who were running from house to house.

The bullets striking the AAV next to them didn't cause either Marine to flinch as they fought. They were both so focused on the fight happening around them that neither noticed it when their Platoon Commander, Second Lieutenant Frank Karsen, came running to their position.

Karsen was something of a legend among many of the Marines he led. Prior to graduating Officer Candidate School, he had won the wrestling national championship twice as a student at the University of Michigan.

"Corporal Queen," Karsen said as soon as he arrived.

"Sir," Queen replied without looking at his Platoon Commander.

"I want you to gather your squad and go kick in the doors of the buildings around us. Kill or capture anyone with weapons and destroy any weapons you come across, even if the building you've found them in is empty," Karsen said.

"Yes, sir," Queen said.

Lieutenant Karsen didn't stick around to make sure Queen carried out his orders. He knew the lance corporal would do so. Having said what he had come to say, Karsen ran off in search of another squad leader to send on a house to house search.

Three Iraqi soldiers watched from the lower floor of one of the houses as Alpha Company disembarked their vehicles and fanned out along the street and adjacent alleys around it. All three of them were conscripts and not a single one had any love for or loyalty to Saddam Hussein. While the American forces had joined the military voluntarily, most of the Iraqis that served Saddam were there simply because they didn't have a choice in the matter.

"I don't like this," one of them said as the U.S. Marines began to pour out of the AAVs and Humvees.

"They were supposed to have just kept driving past us," another Iraqi said, his voice bordering on panicked hysteria. "They weren't supposed to stop. Why are they stopping?"

All three of them had already laid their weapons down and not a single one of them had any desire to pick it back up again. To each of them, the Marines in their Kevlar helmets, NBC suits, all of their gear and with the varied weapons they carried were the most frightening thing they had ever seen.

"They were supposed to have kept going. Why didn't they keep going?" the panicked Iraqi said.

"If they had kept going, we might have stood a chance. Now, with them staying here while we fight, we're dead," the third Iraqi soldier said.

"I don't even believe in Allah," the first Iraqi soldier said.

The three of them watched the firefight raging in the street in front of them for several minutes, while at the same time, doing the best they could to stay out of it. They didn't want to fight and they certainly didn't want to die.

"Surrender?" the third Iraqi asked.

"If we try to surrender, the Americans will kill us," the first Iraqi soldier replied.

"There were plenty of us who surrendered the first time the Americans came and they didn't get killed. That's just something the Saddam's thugs told us to keep us from giving up," the third Iraqi said.

"I'm happy to surrender. We're dead if we don't," the panicked Iraqi soldier said.

They looked at each other in silence for a few, very uncomfortable seconds.

"So … what? Surrender?" the third Iraqi soldier asked.

"Surrender," the first Iraqi agreed. His head lowered in fear, not in shame.

"I'm happy to surrender. I don't want to die here," the panicked Iraqi said.

They saw a squad of Marines beginning to search houses. "Surrender now," the panicked Iraqi said.

At the same time, the three of them picked up their AK-47s and walked to the house's door. They stood off to the side as the first Iraqi opened the door. American bullets filled the open doorway immediately and ricocheted around the house like rabid pinball balls.

When the heavy fire finally died down, the Iraqi soldiers raised their weapons above their heads and, fully expecting to be brought down in a hail of bullets, all three of them walked out of the house.

Immediately, they saw six Marines approaching them, with their M-16s pointed at them, and they heard someone yelling, "Down on the ground! Put your faces on the ground now!" in Arabic.

The three of them didn't waste any time following the instructions to the letter. As soon as they were lying on the ground, they felt Marines pulling their arms behind their backs and putting plastic flex cuffs on their wrists. They were just the first of many prisoners that would be taken that day and of thousands throughout the course of the war.

CHAPTER THIRTEEN

The Marines of Charlie Company didn't have any idea of just how bad their day was about to become as they fought their way through the streets of El Jasiph.

"You all prayed up Private Schuman?" Sergeant Mark Valentine asked.

Twenty-year-old Charles Schuman hadn't enlisted right out of high school, instead he took a couple years to travel around the country and visit with family members he hadn't gotten to see too often while growing up. That is why he was twenty and only a private instead of being a lance corporal, corporal, or possibly even a sergeant.

Schuman wasn't even considering the possibility of a career in the Marine Corps. He was here simply because he had a deep-seated belief that all American men owed it to their freedom, and those who had fought and died over the years, to wear the uniform for at least one term of service. When his enlistment was up, he planned on attending Ohio Valley University, which was a Christian school in West Virginia, and studying to become a missionary. He was devout in his religious beliefs, a fact that caught him a lot of good-natured ribbing from the Marines he served with.

They might have given him a hard time, but most of them held a lot of respect for the fact that Schuman was as religious as he was, but at the same time he didn't throw his beliefs into anyone's face who wasn't interested in hearing about them. Even the atheists in the regiment had a certain degree of respect for him instead of the scorn they held for most religious people. Not that he wouldn't discuss his beliefs with others. In fact, prior to this push into Iraq, he had baptized seventeen of his fellow Marines back at Camp Nicholas. "You bet I am, Sergeant," Schuman replied, slapping his hand down on Valentine's shoulder. "Got my prayers all said, so I'm ready to face whatever comes my way, knowing that I'll wake up in eternal paradise if I breath my last today."

"It sounds pretty ugly out there," Valentine said, with a touch of nervousness in his voice.

"That it does, but there's no sense in worrying about things that we don't have any control over," Schuman replied.

"You've got a point," Valentine said. "I guess I just don't like having to sit on my ass, hearing bullets hitting the outside and knowing all along that any of them could puncture the hull and start bouncing around in here with us. I'd much rather be out there, where I can move around and fire back."

"True enough. Me too, but I'm not too worried about it. Sooner or later, the track is going to stop and we will all be wishing that we were back in here with a little protection between us and the bullets being fired at us."

Valentine looked at him and smiled. "I guess you're right."

Unnoticed by the Marines below, two Iraqi soldiers stood on the roof of one of the houses that Charlie Company was passing. They loaded an RPG into the launcher and one of them rested the weapon on his right shoulder. He took careful aim on one of the tracks that made its way through El Jasiph and pulled the trigger.

The RPG shot out of the launcher and streaked its deadly way towards the Marines below, leaving its distinctive trail of black smoke in its wake.

A sad fact about AAVs is that they were designed to land Marines on the beach during amphibious assaults, like the battles of Iwo Jima and Tarawa during World War II, and not for urban warfare. Their hulls were not armored, which meant if enough bullets hit it, they would start to puncture through and strike the Marines riding inside.

If the hull of a track was questionable in its ability to stand up to enemy bullets, it was worthless to protect against an RPG that hit it. The RPG blew through the track's hull, like a BB through a pop can, landed next to the vehicle's supply of anti-tank rounds and detonated.

The United States Marine Corps had just suffered its first fatalities of Operation Iraqi Freedom.

Lance Corporal Dalton Crocket had been driving the Humvee directly behind the track that had been hit. He watched in horror as the RPG streaked in, on what he knew was a collision course for a direct hit on the track, but he had been helpless to do anything but watch it happen.

His eyes grew wide in fright as he watched the twenty-eight ton vehicle leap into the air when the RPG exploded inside it. The way it jumped, Crocket thought the track kind of resembled the ferret he had had as a child when it was playful. His horror only grew when he saw the bodies of three Marines come flying through one of the top firing hatches before the track landed on the ground again.

One of the Marines that had been thrown out landed on his head, another landed flat on his back and the third landed on his side. Without even realizing he had done so, Crocket pulled his Humvee out of the convoy, shouted, "Cover me!" and jumped from the vehicle.

Knowing Crocket would need help, Corporal Stan Fackler jumped from the front passenger seat and ran with Crocket to the first body, the Marine who had landed headfirst. At some point, the Marine's Kevlar helmet, which was meant to protect against head injuries, had come off and his skull was now split wide open, like an overripe melon. From the vacant, *no-body-is-home* look in the man's eyes, they knew he was dead.

The open skull gave both of them an unobstructed view of the Marine's brain, which was now covered with Iraqi sand. Crocket and Fackler looked at each other. Both of them were in their early twenties and, for the first time, the reality of their situation hit home. This wasn't a game. They were really at war and one of their brother Marines was laying on the ground in front of them, dead. It wasn't one of the simulated deaths they'd seen in training exercises. This Marine, someone they both knew, someone they had trained with, someone they had traded jokes and barbs with, was really dead.

They looked at each other wordlessly. Crocket grabbed the fallen Marine's upper body and Fackler grabbed his lower body and they carried him back to the Humvee. It was full of living Marines, but it was now an ambulance, and those inside would just have to make room for the dead and injured.

After handing the deceased Marine over to those still in the Humvee, Crocket and Fackler ran to the second Marine who had been thrown from the track, ignoring the Iraqi bullets that were coming close to causing them to be included in the day's casualty count. They came to the Marine who had landed flat on is back. He was unconscious and bent at an angle that suggested his back was broken in at least one place, but he was still breathing. Fackler and Crocket picked him up as carefully as they could carried him back to the Humvee.

"Be careful with this one, his back is broken," Fackler said to the Marines in the Humvee. As they ran to the third Marine who had been thrown from the track, Crocket and Fackler both saw the man's body get riddled with bullets from an AK-47. They didn't know if he had survived the explosion or being thrown from the track, but neither of them held any delusions that he had survived being shot up the way that he had been.

The enemy fire was picking up and seemed to be increasingly directed towards them, so the two Marines ran as fast as they could to the body of the man lying on the sandy road. They picked him up, carried him back to the Humvee and gave him to the Marines in the backseat.

With all three Marines, living or otherwise, recovered, they returned to their original seats in the Humvee and Crocket put it into drive and fell back into their position in the convoy.

Crocket was shocked to see the track was still moving forward, despite having been hit with an RPG and having had its antitank ammunition cook off inside it. He would have thought for sure that being hit the way it had would have been the death nail for the track. As the track continued to drive, the gear of the Marines inside it, which had been strapped to its exterior, was in flames and a thick, black, oily smoke poured out of the firing hatches.

My God, Crocket thought as he drove, *I feel bad for those guys, but I'm happy I'm not one of them.*

When he heard the explosion inside the track, Captain Aber turned to look at his First Sergeant, who was operating the radio in Charlie Company's command track.

"Who's hit First Sergeant? Who did we lose?" he asked.

"T-12, sir, but we didn't lose them," First Sergeant Warren Backer replied.

"What do you mean?"

"They were hit by an RPG, which punched through. From what I'm told it looks like their ammunition inside cooked off, but they're still with us, they're still driving," Backer answered.

"Any word on how many we've lost?" Aber asked.

"No, sir. Some of the men were thrown out of the track, but they were picked up by a Humvee. No word on their condition either."

"Shit! It's started." Aber thought.

He moved to where Backer was and said, "Step aside First Sergeant. I've got to let Colonel Everett know we are taking losses."

Charles Schuman regained consciousness after a few minutes. He immediately saw the pitch-blackness that was caused by thick, black oily smoke that covered the entire interior of the track. His first thought was, *What happened? Did I die?*

Then the smoke started to burn his eyes and the screams of pain from those around him reached his ears, so Schuman knew he was still alive. He didn't feel any pain, but knew that didn't necessarily mean he hadn't been wounded.

The first thing he did was start feeling around his own body, feeling for any open wounds, making sure all his limbs and fingers were still where they should be, and so on. Once he was positive that he was still intact and more or less uninjured, Schuman started looking around. Because of the thick smoke, he couldn't see much, but he did see Sergeant Valentine, sitting right where he had been when they were talking just a few seconds before. The only difference was his head was bent almost all the way back. Something, more than likely a large piece of shrapnel, had so thoroughly cut his throat it had even cut through his neck bone. The only thing keeping his head attached to his body was the skin on the back of his neck.

Without even realizing he was coated in Valentine's blood, Schuman started crawling around the track on his hands and knees, feeling around with his hands for the others. From that moment until the time the track came to a stop, his only mission would be doing his best to keep the wounded alive until a corpsman could get to them.

CHAPTER FOURTEEN

Corporal Doug Young led his squad down an unpaved street in advance of the Bravo vehicles that had managed to avoid becoming mired in the sewage trap the Iraqis had set. Back at home, every street had a street sign informing you what its name was. That wasn't a luxury they had here and, while keeping an eagle eye out for any Iraqis carrying weapons, Young was trying to figure out how he was going to relay the path his squad had charted for them back to the new command track. He assumed that he would just have to remember where he turned, which direction he turned and so on because if he fed the wrong information back to the company, they would end up in a much worse situation than being stuck in sewage.

A squad consists of twelve Marines and could be further divided into three fire teams, each made up of four Marines. A fire team consisted of the team leader, who carries an M-16/203; a rifleman, who carries an M-16; an automatic rifleman, who carries a heavy machinegun called a SAW; and the automatic rifleman's "A" gunner, who carries an M-16 as well as all of the ammunition for the SAW. So in Young's squad, there were three SAWs and nine M-16s, one of which had an M-203 grenade launcher mounted underneath its barrel.

Not everyone had an M-203, and quite a few of those who did wished that they didn't since the M-203 added five pounds to a weapon that already weighed almost eight pounds. It was capable of firing a grenade accurately up to fifty meters. It also had its own sights and trigger.

Altogether, that wasn't a shoddy amount of firepower at Young's command, but, at the same time, the twenty-one year old corporal was not thrilled with his assignment to patrol ahead, in advance of the rest of the company. They had been under heavy fire ever since they first entered El Jasiph, and Young was fairly certain those same Iraqi soldiers who had been shooting up their vehicles wouldn't turn up their noses at an opportunity to shoot up Marines patrolling on foot.

They weren't on a pleasure stroll, and the way the squad moved evidenced that. Under different circumstances, they would have moved much faster, but not in a city with thousands of people who wanted to kill them. The squad moved in a staggered column. That meant the Marine in the lead was on the left side of the road, the second Marine was on the right side of the road, the third was on the left side, the fourth was on the right side and so on throughout the entire squad. To avoid falling into any of the hundreds of ambushes that waited for them throughout El Jasiph, the squad moved one man at a time. When they weren't moving, everyone in the squad was on one knee, to make a smaller target of themselves, and had their weapons trained on the courtyards, doorways and windows on the other side of the street. None of them spoke.

Young led his squad from the front and was surprised by a uniformed Iraqi soldier walking out of a courtyard forty feet ahead of him. The man had clearly not been expecting to come across any Americans in this area of the city. His AK-47 was slung over his shoulder, his hands were in his pockets, a cigarette hung from his lips and he moved as if to a musical tune that only he could hear.

Corporal Young had been so thoroughly drilled that his training became instinct in that moment and he put the Iraqi in his weapon's sights. He gave the trigger one quick squeeze and the man's right side seemed to just disappear, leaving a huge gaping and gory wound where it had been.

As soon as the noise he made by firing his weapon hit Young's ears, he wished the Marine Corps issued M-16s with built in silencers. He was certain the sound of a weapon being fired would bring hundreds of Iraqi soldiers running to their position.

The force of the bullets impacting his body threw the Iraqi soldier to the left and he just lay there in the sand. It was the first dead body Doug had ever seen, and it had certainly been the first time he had ever killed a man.

In the immediate aftermath of the shooting, all twelve members of the squad tensed and kept their weapons ready for the Iraqis they were all certain were responding as rapidly as they could. After several anxious minutes passed and no one responded, the Marines breathed a little easier. With all the weapons being fired within the city limits of El Jasiph, Young's must have gone unnoticed. But that left the Marines with one question: how many Iraqi soldiers were there around them? Had they found an area of the city that didn't have a very large enemy presence?

The tension each of the Marines felt didn't lesson any as the squad slowly advanced down the street. They did see a few unarmed Iraqi civilians, and the Iraqis saw the Marines, but so far the civilians hadn't raised any alarms about their presence.

Private Don Bagley, one of the squad's SAW operators, was kneeling when he saw one of the Vietnam era M-60-equipped pickup trucks turn into an alley from an adjoining alley. Even though he was just days shy of his twentieth birthday, Bagley kept calm and didn't waste any time in pointing his heavy machine gun at the pickup truck and squeezing its trigger.

The pickup truck's windshield disappeared under his onslaught. There wasn't any question that its driver had been killed. Like the soldier that Young had killed, this pickup truck and its driver didn't appear to be rushing into the battle. The M-60 in the bed didn't even have an operator with it.

The sound of one of his SAW operators opening fire made Young's heart stop and drop into his stomach.

Young, along with the rest of the squad members in front of Bagley; turned with their weapons ready, to see what he was shooting at. Their ears were met with the sound of a vehicle crashing into a building.

CHAPTER FIFTEEN

I can't get a shot from down here, Staff Sergeant Rich Olan, a sniper assigned to Alpha Company, thought as he tried to get an Iraqi soldier in the cross-hairs of the scope mounted on top of his M40-A3 sniper rifle. Try as he might, Olan just could not get a good enough angle to guarantee a one-shot-one-kill shot.

He lowered his rifle and wiped the sand and sweat from his face before it could get into his eyes. Olan looked over at the sniper next to him, Sergeant Henry Quartermaine, and saw he was having the same trouble acquiring a target.

We have to find a better perch than this, Olan thought as he looked around the street. He saw Iraqi soldiers standing on a rooftop firing down at the American Marines. *That's where we need to be, on a rooftop.*

Olan looked over at Quartermaine, who was looking intently into his scope, but not firing, apparently oblivious to the heat and the sweat running down his face.

"Henry," Olan said.

"Yeah," Quartermaine replied without taking his focus off whatever he was looking at through his scope.

"Any luck?" he asked.

"Nope," Quartermaine answered, still apparently not paying any attention to him.

"What do you say we grab Vince and Rob and get to a better vantage point?"

Quartermaine raised his head from his scope and looked at Olan. "What're you thinking?" he asked.

Olan pointed to a rooftop ten houses down the street from their current position. "I'm liking that rooftop there. It has alleys on both sides of it so we will have an eagle's eye view on them as well as this street. It's also a little taller than the houses around it so we will have an easier time dropping the Hajis on their roofs."

Quartermaine looked at the house Olan had pointed out for a few seconds, with the critically appraising eye of an auction house examining a possibly counterfeit painting.

"Yeah, that'll work nicely. Much better than trying to shoot from down here," Quartermaine said, approval in his voice.

The two snipers stood and moved off to find the other two snipers from Alpha Company, Corporal Vince Watts and Corporal Rob Beverly.

Lance Corporal Barry Queen's twelve-man squad took up position outside a house with an Iraqi on the roof shooting RPGs at Alpha Company. Fortunately, he was a bad shot and Queen didn't intend to give him the chance to improve his aim.

With one of his SAW gunners backing him up, Queen kicked the door open and immediately dove to the right when someone on the other side started shooting at him. He heard the SAW bark out and stop firing.

Oh yeah, Queen thought with his face still in the sand. *Today is really going to suck.*

Even before Queen could get back to his feet, his squad was storming the house. He hurried to fall in behind them.

Inside, his squad broke down into its fire teams and started clearing the house, room by room. As he passed by the corpse of the Iraqi who had shot at him, Queen kicked it in the jaw as hard as he could.

The rest of the first floor of the cinder block house was clear. As they moved up the stairs to the second floor, the squad came under fire again. The Marines returned fire quickly enough that it drove the shooter back and away from the stairs and they hurried the rest of the way up before he could return and shoot at them again.

They didn't see anything at the top of the stairs; the shooter had already beat a hasty retreat. Aware of the fact that there were still at least two hostiles in the house, Queen's squad advanced through the second floor, with their weapons up to their shoulders and ready to be fired at a split second's notice, clearing each room as they went.

Queen kicked open a door and rushed in with the rest of his fire team right behind him. Huddled in a corner were an elderly Iraqi couple who appeared to be in their late seventies or early eighties. Both were clearly terrified and not a threat to him or the Marines under his command.

The wife started screaming as loud as she could and the husband started saying something in Arabic. Even though he couldn't understand the man's words, Queen could tell he was begging him not to hurt him or his wife.

"Clear," Queen said and motioned for his fire team to back out of the room.

Once out of the room, he looked to his rifleman and said, "Stay here. Don't let these geezers leave this room until we've got this place secured. Last thing I want is for them to start running around, like chickens with their heads cut off, and get killed in a crossfire."

"Sure thing boss," the young private first class said and knelt, covering the way they had come from with his M-16. As soon as the rest of the squad finished clearing the floor and moved onto the roof, he'd continuously sweep the entire hallway to make sure no surprises crept up on the squad from the rear.

The sound of an AK-47 being fired filled the hallway and Queen saw one of his men go down. The rest of the fire team immediately fired their weapons into the room.

"Damn!" he yelled as he ran to where his Marine was lying on the ground.

When he got to where his man landed, Queen saw he was alive, dazed, but alive and apparently uninjured.

"Where're you hit?" he asked.

"I've got a headache," the private answered.

"He got hit in the head," a private first class said.

Queen picked the Marine's Kevlar helmet up off of the floor and saw it had two deep dents in it. The young Marine climbed to his feet and Queen tossed his helmet to him.

"Keep it on, it saved your life so it's your lucky charm now," Queen said and motioned for his men to continue clearing the house.

The rest of the second floor was cleared without any further problems. No one went up the stairs to the roof until the rest of the squad was ready, since they knew they would encounter at least one armed Iraqi up there. It didn't matter how many Iraqis were up there; they knew the Marines were coming up and probably had their weapons pointed at the stairwell.

With two of his men backing him up, Queen crept up the stairs with four grenades in his hands. Once at the top, he pulled the pins on the grenades and lobbed them onto the roof in rapid succession. He didn't know where he was throwing them, it was completely random. If he didn't manage to kill every Iraqi soldier on the roof, he at least hoped to discombobulate them enough to give him and his squad a fighting chance at getting all the way up without getting cut down.

After the fourth grenade blew up, Queen pulled his M-16 back up to his shoulder, yelled, "NOW!", and charged up the stairs.

At the top he saw the corpse of an Iraqi who had had the misfortune to have been standing right where one of the grenades had landed. The top half of his body looked like it normally did but the lower half of his body was shredded and in a pool of blood.

Queen immediately began sweeping the rooftop and saw the Iraqi with the RPG launcher. He gave his trigger a quick squeeze and the enemy soldier toppled over backwards, falling from the roof.

He rushed over to where the Iraqi had fallen and looked down. The enemy soldier's body just lay there in the sand below him, bent at impossible angles, broken and lifeless.

It only took a couple of seconds for the Marines to be certain the rest of the roof was clear. As soon as he knew it was safe to let his guard down slightly, Queen looked at his men and said, "Well, boys, one down, only about a thousand more to go."

That brought a round of laughter.

A few minutes later, Queen's squad was in position to take the door of the house next to the one they had just cleared. Queen took the door and this time he wasn't fired at.

They were able to clear the house quickly because no one was there, not even any civilians. What they did find were a bunch of weapons, AK-47s, RPGs, a couple RPG launchers, a mortar tube and a few mortar shells.

"Why the fuck did they leave all these weapons just laying here?" one of the men on his squad asked.

Queen turned his head to look at the Marine. "You ever seen Mel Gibson's movie, *The Patriot?*"

"Yeah, why?"

"Remember the scene, right after one of his sons was killed, when Mel Gibson placed several muskets by trees spread apart from each other. He'd fire one weapon at the British soldiers and move on to where he had the next musket ready, fire it and keep going?" Queen asked.

The man nodded his head, indicating that he remembered the scene.

"Well that's exactly what Haji is doing to us," Queen said.

After a couple of seconds he said, "But we're going to deprive them of these weapons. Check the house, every room, every place a weapon can be hidden and take them all to the roof."

It took less than ten minutes to clear the house of all of the weapons that had been left inside it. Queen had them stacked into a pile and then ordered his men down to the floor below them.

When they were safe, he pulled the pin on another grenade and tossed it on top of the weapons. With only seconds to spare, Queen ran and more dove than anything else onto the stairs. After the explosion, he righted himself and returned to the roof. The pile of weapons had been turned into nothing more than twisted and ruined pieces of metal laying helter skelter across the rooftop.

Queen couldn't help but smile at the thought that none of those weapons would be used to kill any Marines this day. A part of him wished he could be a fly on the wall when the Iraqi soldiers returned to discover the weapons they were counting on being there had been destroyed.

CHAPTER SIXTEEN

Two young Iraqi brothers stood in front of Charlie Company's approaching convoy. Beneath their civilian clothes they both wore Iraqi Army uniforms, they both had AK-47s slung over their shoulders and they both had an RPG launcher resting on their shoulders.

Ever since the Americans had first arrived outside of El Jasiph, the brothers had bragged that they would be the first to kill an American tank. No one ever accused them of being a shining example of the best and brightest that the Iraqi Army had to offer.

The two brothers stood there, excitement pumping through their veins as the Americans approached.

"Ready?" the older brother asked.

"Ready," the younger brother answered.

"Then fire!" the older brother ordered.

They both pulled on their triggers and two columns of black smoke raced away from the brothers. It was all they could see; their RPGs moved too fast for their eyes to be able to see the RPGs themselves.

Both brothers lowered their launchers as soon as their RPGs were away, eagerly waiting to see the grenades they'd fired bring a fiery death to the tank in front of the American convoy.

Instead of striking the tank as the brothers had hoped, the older brother's RPG went wide and traveled down the side of the convoy, hitting and detonating harmlessly on the road. The explosion caught the attention of the Marines in one of the tracks but failed to even hurt a single one of them. The young brother's RPG went really wide and entered a house through its first floor window. Like his brother's, it failed to kill, or even injure, any of the Americans, but his did manage to either kill or severely injure an entire Iraqi mortar team. The younger brother had done the Marines a bigger favor than anyone realized because that team had been waiting for the American convoy to pass so it could begin raining mortars down on it from behind.

The younger brother realized he would have to kill a tank later and ran into the safety of one of the houses down an alley. His brother, however, was determined to kill a tank, but all he was proving is that he was one of the dumbest soldiers in the Iraqi Army. He allowed his RPG launcher to fall to the ground and unslung his AK-47 from his shoulder.

He pulled the weapon up to his shoulder and pulled its trigger, unleashing a torrent of wildly fired bullets, as if those bullets had even a hope of punching through the tank's heavily armored hull and killing the US Marines inside.

Even as the bullets striking the tank bounced away harmlessly, the Iraqi soldier stood his ground. He was determined to own the bragging rights that would come along with killing an American tank.

As the M1-A1 Abram came closer and closer to him, the Iraqi just continued to pull the trigger on his AK-47, sending bullet after bullet ricocheting off the tank as it approached. Even the tank being right in front of him of him didn't deter the Iraqi from firing his weapon. He only stopped when the tank's tread hit him and pulled him beneath the vehicle he had been doing his best to kill. The tank turned him into such a flat wreck of a man that he was barely recognizable as once having been a living person. The other vehicles in the convoy running his corpse over didn't help the condition of his body any.

Charlie Company's convoy continued doggedly on towards its objective on the far side of El Jasiph, even the track that had had an RPG blow up inside it. It wasn't moving quickly. Its useful life was coming to a rapid end and it had to fall back to the rear of the convoy so it didn't hold up any of the fully functioning vehicles, but it followed the rest of them.

CHAPTER SEVENTEEN

First Lieutenant Owings was not a happy man. It wasn't that he disagreed with Captain Earl's decision to take most of the company to support Charlie Company since that was their mission; he just wished his commanding officer would have waited until air cover arrived before taking so much firepower with him. Instead, he had fifty men to protect their mired vehicles with. Owings knew their air cover would be coming soon, but how soon was the question.

It looked like there had to be at least four hundred Iraqis shooting at his fifty Marines and Owings didn't like those odds at all. His men were good, well trained and all acting according to their training, but he was all too aware of the lessons learned in Somalia during the Black Hawk Down incident. A larger military force, even one as bad at shooting as these Iraqis seemed to be, could inflict heavy losses on a smaller military force, no matter how well trained that smaller force is.

So far he hadn't lost anyone, a fact that Owings was grateful for, but he wondered how long it could continue. As he looked around, he saw men lying on the ground returning fire without any cover. He saw men who had advanced to the buildings closest to the flood of sewage so they could take cover behind them while they fought. Others had made small mounds of sand to hide behind and others even hid behind the bodies of dead Iraqi soldiers. Some even stood in the sewage and returned fire from the cover the mired vehicles provided.

Owings was proud of the Marines that were here with him and this was a day that he was looking forward to telling his children and grandchildren about endlessly, when he eventually had children, that is. He hoped that air cover came quickly enough that these Marines would all still be around for him to introduce these children and grandchildren of the future to at the reunions he was sure they would have someday.

Private Larry Rolling stood next to the stuck command track, in knee deep sewage, as he fired at the Iraqis, many of whom had left the buildings they had been hiding in by this point and were making half-hearted charges at the Marines.

"Why are we protecting these things? Why not blow them up and move on with everyone else?" he asked.

Ignoring the splashes that Iraqi bullets were making all around them, Corporal Dan Abraham replied, "Hell if I know. The brass doesn't include us corporals in on that kind of discussions."

Several minutes passed while they fired back at the Iraqi soldiers, breaking yet another advance on their position.

It's almost like they are making a show at overrunning us, but don't really want to, Abraham thought as he pulled the trigger on his M-16. *But then it could be that they just don't want to tangle with our weapons.*

"Do you think we'll be all fucked up from this shit? Like how my uncle came back from Vietnam?" Rolling asked Abraham without looking at him.

"Not me. First thing I'm going to do when I get home is go to a tittie bar. Boobs and beer is all this boy will need to get his head back on right."

A few more minutes went by before either of them said anything else.

"Sucks I have to wait until I'm twenty-one to get my head back on straight," Rolling said. He was laughing.

"Make you a deal. You survive this shit and as soon as we get home I'll rent a hotel room, buy the beer and hire some strippers that do private shows. No putting up with the crowd at a club, I'll hire them for the two of us and some of the other guys," Abraham said as he saw an Iraqi fall, one he thought he had hit.

"Can I get that in writing?" Rolling asked, his voice sounded serious.

"What? Why do you want that?" Abraham asked.

"That way, if you don't make it home, your parents will know that having a hot woman put her tits in my face was your final wish," Rolling answered, laughing again.

Abraham laughed despite everything that was happening around him. He couldn't help it, what Rolling said had been funny. Instead of saying anything in reply, he just stopped firing for a second, smacked Rolling lightly in the back of his head and went back to the business at hand.

A few more minutes went by and Abraham said, "Keep talking like that and you might just discover that the dancer I hire for you is a drag queen."

Rolling laughed again, surprised that he could find levity while bullets were smacking into the track next to him and the splashing in the sewage all around him. "You're cold man, very fucking cold."

The familiar whump, whump, whump of helicopter rotors hit their ears and a couple seconds later, two AH1Z Super Cobra helicopters appeared from within the city. One of them strafed the Iraqis on the street who had been in the middle of another half-hearted charge with its three-barrel twenty millimeter Gatling gun. Those who weren't killed or severely wounded outright fled the streets to the safety of the surrounding buildings.

An Iraqi mortar crew made the mistake of firing another mortar at the Marines just as the helicopters arrived on the scene. The pilot of the second Cobra saw it launch, sighted in on their rooftop position and fired a Hellfire missile. The rooftop disappeared from the building in the same blast that incinerated the mortar tube, all unfired rounds and the three Iraqi soldiers manning the position.

Lieutenant Owings watched the Cobras come in firing and thought, *Hell yeah! That's just what we needed! Now all I need is for that tank retriever to get here so we can get these beasts unstuck and get our assess back on the road.*

The unexpected appearance, and attack, of the helicopters had brought the incoming fire to a sudden stop but the respite was short lived. It quickly picked back up again.

CHAPTER EIGHTEEN

Staff Sergeant Rich Olan watched the chaos that was developing around Alpha Company through the scope of his sniper rifle as he rested it on the edge of the roof.

Corporal Vince Watts stood over Olan's left shoulder looking through his spotting scope. He called out targets for Olan and helped the shooter dial in the correct range, taking into account factors such as wind speed and distance.

Covering an alley where they'd found a lot of Iraqi activity when they had first taken up position on the roof, Corporal Rob Beverly looked through the sniper rifle's scope while Sergeant Quartermaine looked through the spotting scope.

The civilian population of El Jasiph seemed to have one of three reactions to the battle that raged in the streets of their city. One group was terrified and ran wildly through the streets. All too often, this reaction resulted in their getting cut down in the crossfire of the two opposing military forces.

Others, a really surprising amount of them in Olan's opinion, reacted as if it were some grand stage play being acted out on the streets of the city they lived in. It really shouldn't have surprised him, after all, curiosity has always drawn people to gawk at what they've never experienced and don't understand. This kind of reaction didn't just occur in third world countries either; many well-to-do civilians had made an event of watching the Battle of Bull Run, the first battle of the American Civil War. It hadn't just been men drawn to it either. They had brought their wives and children and many had dressed in their Sunday best for their day of battle watching. They had brought a lunch, and some even brought alcoholic beverages just to add pleasure to what they planned on being a relaxing and enjoyable day. Like the civilians of El Jasiph, those Americans had learned the hard way that war wasn't a spectator sport when the battle turned in direction and came right towards them, causing the noncombatants to run for their lives.

Those who didn't fall into the first two categories seemed to be oblivious to the fact that there was even a battle raging in their city. They just went about their daily business as if they couldn't hear the screams of pain from the wounded and dying men from both sides, as if they couldn't hear the bullets, grenades and mortar shells as they tore buildings and human flesh wide open. These Iraqis just carried on, doing their laundry in the open, shopping and walking with their children as if it was just another day in the city of El Jasiph.

Dumbass! Olan thought as he observed a family of five standing on the second floor balcony of a house. *Get your wife and kids back inside while you still have them!*

Olan found himself having to turn his head when an Iraqi mortar shell scored a direct hit on the balcony. The family disappeared briefly in a cloud of dust and gore as the shell exploded and the balcony they had been standing on fell apart. When all was said and done, five charred, blistered and dismembered corpses lay amid the rubble that had been the mud brick balcony on the sandy street of El Jasiph.

He didn't feel bad for the parents. It was their fault that they had decided to watch a battle, a real battle with real people being maimed and killed, as if it weren't anything more serious than an Arnold Schwarzenegger film from the 80's. He hoped they burned in Hell for their stupidity. The ones he felt bad for were the three young children who had been standing on the balcony with them. They hadn't done anything more than trust their parents to do what was best for them and now they were dead because of the faith and confidence they had placed in their parents. Olan hoped the little ones would get their angel wings quickly.

But, unless he wanted to join that family in death, this was not the place to dwell on such things and he quickly started searching for another target through his scope. He saw an Iraqi woman leave a house pushing a baby carriage.

What's wrong with these people? Olan thought. *Isn't there one of these people who would like to see their kids grow to adulthood and have kids of their own?*

Still, there was something about the woman that smelt fishy so Olan kept his eye on her. He watched as she walked behind a cinder block courtyard wall. A few seconds later, she walked out from behind the wall, still pushing the baby carriage, and back to her house. Within seconds of the woman returning to her house, a mortar was fired from behind the wall.

What the hell? Olan thought. There was clearly a mortar position behind that wall. What he couldn't be sure of was if she had been using the baby carriage to transport a mortar round to the position or if her husband just happened to be part of that mortar position and she had simply been taking their baby to see its father. That would have been innocent enough, dumb, but innocent.

"Building with the blue door, Haji sniper on the second floor, third window from the left," Olan heard Watts say.

"Just a second, I might have someone feeding a mortar position," he replied.

Just as he had expected, Olan saw the woman come back out of the building a little over a minute later pushing her baby carriage. Once again she went behind the courtyard wall, came back out a few seconds later and then another mortar round was fired from behind the wall.

Fucking bitch is using the baby carriage to conceal mortar rounds as she takes them to the guys behind the wall. If I see her again, I'm taking her out, Olan thought.

Sure enough, she came back out of the house a little over a minute later, again pushing the baby carriage. The way her body was positioned severely limited Olan's firing options, so he sighted in on the side of her head.

My God, I hope I'm right on this, Olan thought and pulled the trigger.

The woman jerked to the right as Olan's bullet entered the left side of her head, just above her ear. The entrance wound wasn't any bigger than the bullet itself but the exit wound was the size of a human fist, which gave the mud brick wall next to her a fresh coating of blood, brain matter and skull fragments as her body collapsed to the ground. On its way down, her body toppled the baby carriage and a mortar round rolled out of it, not a baby.

Olan returned his attention to the courtyard wall that concealed the Iraqi mortar position and saw a man, wearing civilian clothes, emerge, firing an AK-47 wildly as he ran for the round that had rolled out of the baby carriage.

The man's body position made targeting much easier than the woman's had. Olan was able to sight in on the sniper's triangle, an imaginary triangle that ran from nipple to nipple, then from one of the nipples up to the soft spot at the bottom of the man's throat and then back to the other nipple. A hit anywhere in this triangle was a guaranteed kill shot. He pulled the trigger of his sniper rifle and the bullet hit the Iraqi soldier in his upper chest. The man fell to the ground and didn't move again.

"Ok, Watts, where's this Haji sniper at?" he asked.

CHAPTER NINETEEN

As his squad moved slowly through the unpaved and sandy streets of El Jasiph, Corporal Doug Young kept each of his senses alert for anything that might represent a danger to him or his men. They advanced at the same slow and deliberate pace they had ever since Gunnery Sergeant Gerold Schuring had ordered him to gather his squad and take them out to recon a trap-free path for the Bravo Company vehicles that hadn't fallen into the sewage quagmire.

He didn't like advancing on foot through a city with thousands of enemy soldiers in it, but then the recruiter never told him that Marines only had to do the things they liked or wanted to do. Being a Marine meant adapting to and overcoming any obstacles that stood in the way of completing the mission you were handed and that is exactly what Young intended to do. It might have sounded like a grandiose delusion to say out loud, but who knows, maybe the outcome of the entire battle of El Jasiph would depend on his ability to successfully complete his mission, like it or not.

His ears picked up the sound of several men talking in Arabic. Even though he couldn't understand what they were saying, anything spoken in Arabic at that point in time sounded decidedly unfriendly. Young motioned for his squad to come to a stop and listened for where the voices where coming from.

It took a few seconds, but Young thought the voices were coming from an open window about twenty feet ahead of him. He motioned for his squad to hold their positions and moved forward on his hands and knees to avoid being seen. When Young made it to the window, he cautiously rose up and looked through it. Inside he saw ten Iraqi soldiers preparing weapons of all kinds.

Young didn't know if they were preparing to move up to the front lines of the battle, or if they were just readying the weapons to resupply the Iraqi soldiers already engaged. What he did know is that he had to take those weapons out before they were used to kill his fellow Americans.

Young crawled back to the rest of his squad and motioned for them to come to him. As soon as they were close enough to allow him to speak quietly, Young said, "Ok, here's the score. We've got about ten Hajis preparing weapons, AKs, mortars, RPGs, the whole works. Those can't be allowed to reach the fight. I'll use my 203 and launch a grenade through the window. Bagley and Evans, you two will be on either side of the window. As soon as the grenade blows, I want the both of you to pop up and flood the room with bullets, pretend you're part of the Saint Valentine's Day Massacre. No one survives. Any questions?"

Both of the privates shook their heads, indicating they didn't have any questions and Young turned to the other nine men of his squad. "Okay, the rest of you, cover that door. I don't want to get tied down with searching this house so you'll just wait to see if anyone comes out the door. If they do, I want them running head-on into a wall of bullets. Understand me?"

Everyone nodded their understanding and the squad broke up and went to their positions. The best firing position Young could find at such close range was sitting on the ground with his back against the courtyard wall across the street. He placed his M203's sights on the window, pulled the trigger and heard the thunk of the grenade being fired.

The grenade sailed through the window, with the apparent effortlessness of a basketball soaring through the air and landing through the hoop. Its arrival was met by several started yells in Arabic and then, seconds later, an explosion that sent dust and chunks of cinder block flying out the window.

Exactly as they had been ordered too, Privates Bagley and Evans popped up and started filling the room with fast moving bullets. After a few seconds, they stopped firing and looked into the room. It was a bloody mess and nothing moved.

Those waiting to ambush anyone who came through the door waited tensely for several seconds. When no one came out, Young decided they had successfully cleared the building of Iraqi soldiers and ordered his squad to continue their patrol through the city.

They patrolled for another twenty minutes before the sound of heavy vehicles moving down the street they were on reached their ears.

Please let those be ours, Young thought, but ordered, "Take cover! Take cover!"

His squad did exactly as he had ordered and seconds later, the vehicles of Bravo company that hadn't gotten mired in the sewage rolled through. The driver of one of the Humvees saw Young and pulled over.

"Come on boys, jump in. I ain't got all day ya know," the driver said.

Young's squad hurriedly climbed aboard the Humvee and other vehicles that had stopped to pick them up.

Sitting in the backseat of the Humvee, Lance Corporal Doug Young leaned back in his seat and thought, *Thank God that's over.*

The appearance of the American vehicles brought them under heavy fire immediately and Young's men joined the rest of the Marines in firing back as they pushed their way through El Jasiph to assist Charlie Company in their advance.

CHAPTER TWENTY

Private First Class Ron Farber still manned the Humvee's fifty caliber. All around him the Humvee was scarred and dented from being struck by Iraqi bullets, but so far none of the thousands, if not more, of bullets that had been fired his way had hit him. It was a fact that Farber was grateful for and on a subconscious level he prayed that his luck would continue.

He had fired thousands of rounds back at them and, if his memory served him correctly, had sent around thirty Iraqis to shake hands with Mohammed and Allah face to face. From the time he found out that he was being deployed to Iraq, Farber had wondered how well he would be able to handle killing other people. Only the future would be able to tell how he reacted to it down the road, but now, in the heat of the moment, it wasn't bothering him in the least.

In the movies Farber had seen about World War Two and Vietnam, there was always one person who went through a great deal of soul searching after killing someone. That person always went through the enemy soldier's wallet and cried over the pictures of the dead man with his happy family and the fact that he wouldn't be returning to that family. Farber wasn't soul searching, he didn't have any desire to go through anyone's wallet, he didn't care if he was widowing any wives or orphaning any children because each of the people he had killed wouldn't have thought twice about killing him or one of his brother Marines.

As much as he hated sitting in such an exposed position while riding through a city where everyone seemed to be determined to kill him, he was really happy that he wasn't the poor soul who was stuck driving the lead vehicle. As a child, Farber had worked his way through maze puzzles and he had the distinct impression that the people who had designed the street layout for El Jasiph had been determined to make the city more confusing than the maze puzzles were, and those were tough.

Even though he didn't like standing in such an exposed position, being on his way somewhere was better than holding tight in one location. Farber had really become uncomfortable when the vehicles in the lead had gotten stuck in the sewage and brought the entire company to a stop. Just because their forward momentum had come to a stop didn't mean that the incoming fire had. If anything, it only seemed to increase, but Farber realized the increase of hostile fire could have just as easily have been his imagination. There was one thing he knew for sure though: he felt bad for the guys who were left behind to babysit the stuck vehicles. That wasn't a duty he would have wished on anyone.

Did I make the right call? Captain Earl second-guessed himself. He was so wrapped up in his self-questioning that he no longer even noticed the pinging sound of Iraqi bullets bouncing off the AAV's hull or the occasional bullets that punctured through its hull and bounced around inside the vehicle. He didn't even notice the horrible smell that had been drifting off of him ever since he went wading through the Iraqi sewage. *Should I have left those men behind? Should I have left anyone behind, or should I have left more behind? Did I leave enough men for them to make their stand or did I just waste fifty lives? Should I have just abandoned those vehicles and kept the company intact?*

The words, "Holy shit!" snapped Earl out of his thoughts.

"What? What is it?" he asked.

"How did you miss that, sir?" the track commander asked.

"Miss what?"

"We just had a mortar hit the road no more than fifteen feet in front of us, sir," the track commander answered.

Earl shook his head. "There is a Hell after all and the Marine Corps has dropped us right into the middle of it," he mumbled under his breath.

The Company Commander bent over and looked out the front of the track. Ahead of them lay another wide-open space that looked to be about the same size as the one the Iraqis had flooded with sewage.

I don't know if leaving those guys back there all by themselves was the right call or not, but I really don't like the idea of making them fight their way through these streets on their own, he thought.

"Get us out into the center of that open area and bring us to a stop," Earl told the driver.

He turned to First Sergeant Warren Backer, his company's first sergeant, and said, "Get on the horn. Tell everyone who isn't manning a top-mounted weapon that they'll be disembarking as soon as we come to a stop. I want a three hundred and sixty degree defensive perimeter set up immediately."

"Yes, sir," First Sergeant Backer replied.

I just hope that Charlie Company can make it without our help, Earl thought.

A few minutes later the track came to a stop and Earl picked up his M-16. He had ordered all his men to disembark and he was going to be right there fighting alongside them.

The rear ramp lowered and everyone who wasn't manning either the fifty caliber or the TOW top turrets charged out into what they expected to be a veritable thunderstorm of Iraqi bullets. Instead of heavy enemy fire, the Marines encountered nothing at all as they exited the vehicles that had brought them there. The Iraqi soldiers had completely quit firing on them.

It was a development that confused Earl, but he wasn't complaining. His men could use a few minutes to catch their breaths. However, just because they weren't under fire now didn't mean things would remain so calm and he went around telling his Platoon Commanders where he wanted their platoons positioned in order to establish the three hundred and sixty degree defensive perimeter that he wanted established.

The space they were in seemed to be an oasis within the labyrinth of El Jasiph. Earl wanted the perimeter up as soon as it could be put into place because the Iraqis were bound to start firing on them again. He wouldn't have even put it past them to charge and try to overrun their position.

Shortly after the perimeter was in place, First Sergeant Backer came running up to Earl. "Sir, you need to come over to Third Platoon and see what's coming," he said.

"Iraqis?" Earl asked, instantly worried about a charge.

"Yes, sir, just not in the way you're thinking. We've got civilians walking towards us," Backer answered.

"What the hell…" Earl said, but walked away from Backer and towards Third Platoon before Bravo's first sergeant could say anything else. Backer was right behind him.

When he arrived to Third Platoon, Earl saw what had to have been close to a hundred civilians approaching. Every now and then he would hear one of them say, "food" or "doctor" in English.

He looked at Third Platoon's Platoon Commander and said, "Do not let them come close to us, Lieutenant."

"Sir?" the first lieutenant replied.

"Haji has been dressing their soldiers like civilians all day to fuck with our minds," Earl replied. "If even a couple of them are soldiers carrying grenades, just imagine how bad things could get here. Or, even worse, what if they've taken a page out of the VC's play book and have explosives strapped to those civilians. Just imagine the mess that would make of us."

The VC he was referring to were the Vietcong. The VC were a political and military force from North Vietnam and Cambodia that fought against the American and South Vietnamese forces using guerrilla and regular army units. Quite often their guerrilla tactics involved strapping bombs to children and then sending them up to American servicemen to ask for candy. Once the child was where they needed to be, the bomb would be detonated. The resulting explosion would kill both the child and the American.

"Yes, sir, I see your point. How do you want us to keep them back?" the first lieutenant said.

"Have your men fire well above their heads and into the ground far in front of them. I don't want any civilians accidentally being hit," Earl answered.

"And if some of them don't turn back, sir?" the lieutenant asked.

"If they don't take the hint, then drop their asses. If they keep coming after our warning shots, then they've probably don't have anything good in mind," Earl answered. The platoon commander didn't wait for further orders. Instead he turned away from his company commander and began relaying their orders to his men.

Earl turned to Backer. "I want you to give those exact same orders to the other platoon commanders. I don't want anyone who isn't a U.S. Marine getting through to us," he said.

"Yes, sir. Right away," the company's first sergeant said and jogged away.

He turned his attention back to the approaching civilians just as Third Platoon opened fire with their warning shots. Just as he hoped they would, the civilians scattered and ran back towards the city.

Thank God, Earl thought. *The last thing I want on my conscience is giving the order to gun down civilians who are only trying to get food or medical attention. I hope that tank retriever gets those stuck vehicles unstuck soon so we can keep moving.*

CHAPTER TWENTY-ONE

Charlie Company had had a rough time of it. They were worn, battered and one of their tracks, the one that had had an RPG blow up inside of it, was well beyond being on its last leg, but, in true Marine Corps fashion, it pushed on anyway.

Despite having run the gauntlet, they had managed to be the first company to emerge on the other side of El Jasiph. It wasn't a race, however, and as soon as their vehicles came to a stop, the Marines debarked and rushed to set up a defensive position facing the city. They may have made it through the city but they weren't out of the woods yet. There were still plenty of hours left in the day for things to go wrong and the incoming fire hadn't slacked any. The incoming fire was mostly bullets and mortars. The RPGs weren't so heavy now that they'd escaped the city limits of El Jasiph.

The driver of T-12, the track leaving a trail of thick, black smoke in its wake, realized his vehicle was at risk of blowing up so he brought it to a stop well away from the others and immediately lowered the rear ramp.

One of the Marines on board, a private who hadn't been out of recruit training for long at all before receiving his deployment orders to Iraq, was the first to burst forth out of T-12.

"Gotta get the fuck outta here! Gotta get the fuck outta here! Gotta get the fuck outta here!" the private yelled repeatedly as he ran around in panic.

He went down a few seconds later with three Iraqi bullets in his abdomen. His screams of pain brought the company's corpsman running. A corpsman is a combat-trained Navy medic that goes into battle with the Marines to tend to the wounded. Once to the private, Petty Officer Third Class Craig Trevino took his large, backpack-like medical kit off his back and went to work trying to save the young man's life. Inside the medical kit was everything that he could possibly need to keep Marines alive under combat conditions. The nine-millimeter Beretta he carried remained in its holster on his hip and would only be pulled out if it became completely and absolutely necessary for him to do so. He didn't consider himself a war fighter. He was in El Jasiph to save lives, pure and simple.

Trevino didn't have any intention of becoming a career Navy Man. Instead, once his current enlistment was completed, he planned on taking an honorable discharge and going back to school, and eventually becoming Craig Trevino, M.D. He had enlisted in the Navy instead of going to college immediately after high school because he thought the time spent as a corpsman would give him a leg up over other newly minted doctors when he was looking for a job once he graduated medical school. His original plan had been to become a pediatrician since he loved children, but the time he had spent in El Jasiph had given him cause to reconsider. The bullets zipping through the air, the mortars landing and exploding in the sand and the occasional black smoke trails of RPGs passing over his head—and the fact that he had more wounded Marines that he had ever expected to have to deal with—caused his adrenaline to pump. While he didn't love having the angel of death looking over his shoulder, Trevino did love the rush and had decided that instead of becoming a Pediatrician, he could become an Emergency Room Physician. Other Marines were in flames when they came off the track. Those from other vehicles quickly tackled the burning men and used their bodies and sand to smother the flames. Charlie Company's Three Stooges, Privates Burt Farrow, Andy Irvin and Manny Alverez ran into the stricken track. They were all fully aware of the personal danger doing so placed them in, but each of them disregarded it at the moment.

Once inside, the smoke stung their eyes and filled their lungs, which caused them to cough uncontrollably. Each of them grabbed the nearest man to them, without worrying about whether that man was alive, and put them on their shoulders.

Loaded with dead and wounded, the three young Marines ran back out of the track. As soon as they were outside, another Marine was pointing and saying, "Take them over there! They're collecting the wounded over there!"

"Where?" Irvin asked, not seeing where the man was pointing.

"Over there! Over there! Behind that sand berm," the Marine said and then disappeared into the track himself to bring someone out.

The Three Stooges ran as fast as they could, considering they were running in sand, in full combat gear and carrying a fully-grown men on their shoulders. The run wasn't easy, especially when they had to run up the five foot sand berm to deliver those they were carrying on the other side.

By the time the three Marines laid the men they were carrying down on the ground, their mouths and throats were as parched and sore as they'd ever been in their lives. None of them wanted to do anything more than lean with their backs against the sand berm and down a full canteen of warm water, but there were still too many men in the crippled track who were hurt, dying or already dead who were unable to get themselves to safety. Instead of giving in to their desire to relax and seek comfort, the three of them ran back over the sand berm, through heavy enemy fire, and back towards the track. Personal comfort and desires would have to come later.

They passed the Marine who had pointed out the casualty collection point to them as they ran. He had a man over his shoulders whose face was so heavily damaged it would have looked more at home on the body of an alien from a science fiction movie than on a human being. He didn't look like he was among the living any longer, but then none of them had taken the time to check to see if the Marine they had carried was alive or dead either. It didn't matter.

The aim of the Iraqis firing the mortars hadn't seemed to improve any because they soared high over the heads of the Marines and landed far behind their position. However, the aim of the Iraqis with the AK-47s seemed to have improved greatly since the Americans had first entered El Jasiph and little geysers of sand kicked up around the feet of The Three Stooges as they made their return trip to the stricken track.

When they made it back to the track, the three of them entered and grabbed the nearest body to them. Alverez grabbed the corpse of Sergeant Valentine and, due to the thick, black smoke still in the vehicle, he didn't realize how precariously the sergeant's head was hanging on to the rest of his body. He lifted the corpse and the little bit of skin holding the head to the rest of him ripped and Sergeant Valentine's head dropped back to the floor of the track. Alverez didn't realize that he had left part of his man behind, otherwise he would have stopped to grab the sergeant's head before leaving.

On the run back to the casualty collection point, they saw the other Marine running towards them on his return trip to the track. A lucky mortar round chose that moment to strike a pole holding up power lines. The pole fell and the power lines landed on the ground. One of them hit the Marine who had pointed out the casualty collection point to them. His body made spastic jerking motions, which kind of resembled a thirteen-year-old at a middle school dance, as sparks from the live wire flashed all around him, then he fell dead. No one was foolish enough to try to get his body to the casualty collection point, not while those power lines were still live.

Bullets kicked up sand all around them, so The Three Stooges didn't have the time to think about the marine they had seen die. Instead, they ran around the downed power line at a safe distance and carried their loads to the casualty collection point.

Most of those who weren't involved in trying to rescue their brother Marines from the crippled track had taken cover in a four foot trench and were actively returning fire at the Iraqis who were shooting at them.

CHAPTER TWENTY-TWO

Staff Sergeant Dave Ligget and Sergeant Duane Owen lay on a sand berm in the open area where the part of Bravo Company that was still mobile had come to a stop to wait for the others. They were just two of several Marines who lay at the top of the six foot berm with their weapons to their shoulders.

Unlike their Iraqi opposites, who were firing their AK-47s wildly and without any form of discipline, the two non-commissioned officers took the time to aim their weapons and to fire controlled bursts. There wasn't any sense in wasting any ammunition they didn't need to.

They felt good about their position because the sand berm was thick enough that it completely stopped the vast majority of the bullets that the Iraqis fired at them. The few that did manage to punch through were slowed down enough that if they had managed to strike any of the Marines, the wounds would not have been likely to be fatal, or even serious. That didn't mean that they still didn't have to watch out for the bullets that were fired above the berm. There were plenty of those buzzing by the Marines' heads, like angry little, metal insects.

An Iraqi bullet stuck the very top of the sand berm, right in front of Owen's face, sending up a small explosion of sand.

"Shit!" the sergeant screamed. "Oh, fuck, that hurts!"

"You hit?" Ligget asked, concern filling his voice.

"No!" Owen answered, annoyance in his voice now instead of pain. "Fuck no, I ain't hit."

Owen started rubbing both of his eyes vigorously. "Got sand in my eyes is all."

"Fall back and find Doc Smith. I'm sure he has eyewash, water or something to clean them out," Ligget replied, referring to Petty Officer Third Class Kurt Smith, the Navy Corpsman assigned to Bravo Company.

"I can't see to shoot worth dick!" Owen said, as if he hadn't heard Ligget.

"I said fall back and find Doc Smith," Ligget said, more forcefully this time.

"Yeah, I think that's a good idea," Owen said and then started moving away, still rubbing his eyes.

Corporal Pat Nolan was on the same sand berm as Ligget and Owen, only further down, with several Marines separating them. Despite all the training and exercises he had been through, keeping himself hydrated by drinking from his canteen had completely slipped his mind in the heat of battle.

He was feeling the effects of that oversight. He was confused and, despite the stress of fighting in the Iraqi desert, Nolan was not sweating. He was also breathing rapidly and had a dangerously high fever. Under normal circumstances, one of the others would have noticed he wasn't behaving like himself, recognized the symptoms and had him drinking water long before his dehydration had gotten as severe as it was. Even the warm water from their canteens would have been better than nothing at all. However, with Iraqi bullets, mortars and the occasional RPG coming in at them, everyone had other things to pay attention to instead of Pat Nolan.

No longer aware of anything, including the fact that he was in a battle, Nolan laid his M-16 down on the sand and rolled over onto his back. His eyes stared up at the clouds, fascinated by what they saw.

From a young age, Nolan had loved to read about naval battles, fiction as well as non-fiction, pirates, military—as long as it was a naval battle, nothing else mattered to him. It had excited him to no end when he was ten years old and learned that the U.S. Navy had its own Army, this thing called the Marines. Many of the men around him enlisted in the Marine Corps with dreams of earning honor and a chest full of medals in land battle, but not Pat Nolan. He had enlisted with fantasies of boarding enemy ships at sea, via a gangplank of course, and repelling enemy Marines who tried to board the ship he was on. Nolan had been really disappointed when he had learned the truth and kicked himself in the rear for never having asked his recruiter about it.

Very strong winds way up in the troposphere pushed the clouds along at a fast pace. Almost everyone has played the game on lazy summer days where you lay down on the grass and make up things the clouds look like. When Nolan played it now, he wasn't just making up what the clouds looked like; he saw ships. He saw Viking ships, pirate ships and modern aircraft carriers. He was dangerously close to losing consciousness.

Gunnery Sergeant Schuring saw Nolan lounging on the Iraqi sand, as if it were the grass-filled lawn of the house he had grown up in.

"Corporal!" Schuring shouted, to be heard over the sounds of the battle that raged all around them.

Nolan didn't reply. He just lay there looking up at the clouds.

"Goddamnit, Corporal Nolan!" Schuring barked. "If you don't answer me now I'm going to kick you in the ass so hard you'll land in Detroit!"

When Nolan didn't respond to his threat, Schuring started to worry.

"Doc! Doc!" he called.

The corpsman came running. "Yes, Gunny."

"I think Nolan is having trouble with the heat."

All Smith needed was one look to concur with Gunnery Sergeant Schuring's diagnosis. He looked at Schuring. "Gunny, we need to get him out of the sun or he's going to die. I need your help carrying him to a track so I can get an IV started on him."

Schuring didn't say a word. Instead, he slung his M-16 over his shoulder and helped the Corpsman pick Nolan up. The two of them ran as fast as they could, through heavy Iraqi fire, carrying Nolan, who was just a thread's width from death, to the nearest track.

As they ran, an Iraqi bullet hit Schuring in the back of his right calve, completely removing his Achilles tendon from his body. The Gunny dropped Nolan and landed face first in the sand, breaking his nose.

Unexpectedly carrying all of Nolan's weight was more than Smith had been prepared for and he fell as well. Nolan didn't even seem to realize he had even been dropped.

"What the hell!" Smith shouted as he scrambled around on his hands and knees to look at Schuring. His anger disappeared as soon as he saw the gunnery sergeant's condition.

Smith hurriedly pulled a tourniquet out of his medical kit and applied it to Schuring's leg, just below his knee.

"I'll be right back for you Gunny, but I've got to get Nolan out of the sun or he isn't going to make it," he said.

Schuring was in too much pain to speak. It was taking every ounce of discipline he had not to scream and cry like a newborn baby, but he was able to nod his understanding.

Smith picked Nolan up in a fireman's carry and carried him over his shoulders the rest of the way to the track.

CHAPTER TWENTY-THREE

The Battalion TAC had fallen into chaos as Marines from all over El Jasiph tried to report what was happening in their areas of responsibility. It was so bad the Battalion TAC sounded more like an incomprehensible cluster of voices belonging to several people all talking at the same time and trying to be heard over the others in a small room than it did a military radio frequency. Everyone talking was trying to over-talk everyone else and trying to make sense of anything being said was a futile effort at best.

Lieutenant Colonel Everett had not been able to make heads or tails of what he was hearing over the Battalion TAC and had given up trying. He had assigned his driver, Lance Corporal Jack Tays, the responsibility of keeping an ear on the TAC while he joined the fight, along with his Battalion Gunner and Battalion Sergeant Major.

There were several things Everett hated about the way Hollywood depicted his beloved Corps, but his biggest pet peeve was when they showed Lieutenant Colonels as desk jockeys sitting in the Pentagon. Sure, he had a desk back in North Carolina, but whenever the Marines under his command were out on exercises, so was he. And he certainly wasn't sitting at a desk now, he was kneeling and using the front of his command Humvee as cover from the same Iraqi bullets that were being fired at the men of Alpha Company.

Everett fired at Iraqi soldiers as they ran from building to building and courtyard to courtyard. He had killed several, but he wasn't keeping count and would never wish to know the number of Iraqis he killed. It was something he did because it was his duty, not something he did out of enjoyment.

The noise of battle around him was deafening, but even that wasn't enough to hide the noise made by several heavy machine guns opening up at one time.

"What the hell?" he asked, his voice filled more with confusion than anger or fear. "Are those ours?" He thought perhaps Alpha's SAW operators had all opened up at the same time.

His question hadn't been to anyone in particular and he hadn't asked it loud enough for anyone to have heard him. Everett stopped firing his M-16 and let his eyes scan the battleground. It took a few seconds but he saw muzzle flashes that were unmistakably coming from heavy machine guns coming from four different windows in one building.

He was positive the Iraqis didn't have anything as advanced as SAWs, so he assumed they were firing RPKs or old M-60s. Regardless though, those weapons could still do a lot of damage to his men.

Fucking shit! Everett thought. *I've got to have a squad clear that building before those guns chew us up!*

The Battalion Commander wasn't the only one to notice the heavy machine gun fire coming from that building. So did the fifty-caliber gunner on the rear M1A1 Abram tank in the front of the column.

The tightness of the streets had, for the most part, kept the tanks from being able to fire with their main turrets but the fifty-caliber gunners on top of the three tanks had been keeping their weapons busy pouring bullets into the surrounding buildings.

After seeing the heavy machine gun fire begin coming from the building, the rear tank's gunner said something to the driver through the microphone built into his helmet. A few seconds later, the top turret began to turn to face the building with the heavy machine guns. Once it was facing the correct building, a few more seconds passed while those inside the M1A1 aimed their main weapon. Then it fired.

Lieutenant Colonel Everett saw the one hundred-and-twenty-millimeter round strike the building right where the floor of the second story would have been. His heart was filled with anticipatory excitement as the tank's shell buried deeper into the building and then exploded. The explosion turned the building into a collapsing pile of rubble, silencing the heavy machine guns as the building fell and the Iraqis who had moments before been pouring a heavy rain of deadly bullets on the U.S. Marines were crushed to death.

Inside another building, twelve Iraqi soldiers, all with RPG launchers, saw the tank demolish their heavy machine gun building.

"We've got to take out those tanks!" one of them yelled, his voice on the edge of hysteria.

"How?" another asked. He had an RPG launcher too.

The first Iraqi soldier looked out the window silently for a few seconds, weighing his options. He was Fedayeen and in charge of all the activity that took place in that house. He turned to a soldier who was armed only with an AK-47.

"Go to the roof and downstairs, tell the others to fire their RPGs at that tank," the Fedayeen soldier said. Up until that point, none of the men in the house had been coordinating fire with each other on any particular target. Instead, they had been firing at whatever target presented itself to each individual soldier.

He waited a few minutes to give the soldier time to relay his orders to the soldiers armed with RPG launchers on the roof and first floor. After he'd given the soldier enough time, the Fedayeen soldier pulled his RPG up to his shoulder, placed the American tank in his sights and pulled the trigger.

His RPG raced away from the building, making its way with deadly intent to the M1A1 Abram that had just brought down another building. The smoke trail of his RPG was quickly followed by smoke trails of eleven more.

All twelve RPGs hit the tank at pretty much the same time and the tank quickly disappeared in a cloud of smoke and flame. The sight caused shouts of celebratory joy to go up in the house.

When the cloud around the tank disappeared, they saw it was still there and that the only damage it had taken was a few scorch marks on its hull from the exploding grenades and nothing more. Even the fifty-caliber machine gunner on top had had time to drop back inside and seal the hatch before they had hit. The celebratory shouts died just as quickly as they had started.

The Fedayeen soldier just stared dumbly at the tank as its one hundred and twenty millimeter main gun turned slightly and sighted on the building he was in. When it fired he only had time to open his eyes wide and allow his jaw to fall slack as the shell raced towards the building in response to their RPG attack. The tank's shell hit the building and exploded. Those who hadn't been killed in the explosion were killed when the building collapsed into a pile of rubble.

CHAPTER TWENTY-FOUR

Lance Corporal Kevin Hale, of Bravo Company, lay on the unpaved Iraqi street right in front of the sewage trap. Hale had taken cover behind the corpses of three Iraqi soldiers he had piled on top of each other. He didn't think it was an ideal situation by any means, but he was doing his best to stay alive under the circumstances he found himself in.

Hale hadn't known what to expect when he joined the Marines, but he had never imagined, if he actually had to go to war, that he would be fighting in front of an overflowed sewer, with the stench of urine and feces constantly bombarding his sense of smell; he hadn't expected that he would be using the bodies of dead enemy soldiers to try to shield himself from incoming enemy bullets; and Hale had certainly never expected to find himself having to continuously swat away the flies that were drawn by the stench of sewage, blood and death.

In short, Hale was proud to be a Marine, proud to be fighting for America, but definitely not having the exciting adventure Hollywood had spent so many years promising him.

Sergeant Scott Dove lay on the ground, about ten feet from Hale, behind a small barrier of sand he had hastily constructed to stop bullets from hitting him. Unlike Hale, Dove seemed to be actually enjoying himself.

"One little, two little, three little dead Iraqis," the sergeant sang as he fired his M-16. "Four little, five little, six little dead Iraqis."

Hale looked over at Dove and saw what could only be described as a playful gleam in his eyes as he fired his weapon.

"You sound way to happy over there!" Hale hollered to be heard over the sounds of the battle raging around them.

"Hell yeah I am. I'm happier than a fat kid in a swimming pool of pistachio pudding," Dove replied.

"You've gotta be shittin' me," Hale said, his voice heavy with disbelief.

"Hell no I'm not. I was born for this shit," Dove replied.

A few seconds of silence passed between them as they both fired their weapons at the Iraqi soldiers.

"Maybe if I weren't hiding behind dead bodies, I'd be enjoying myself more than I am!" Hale shouted.

"You ever hear the story of the Three Little Pigs?" Dove asked.

"What's that got to do with anything?" Hale replied.

"Just answer me goddamnit! Have you ever heard the story of the Three Little Pigs?"

"Yeah, sure, when I was little. Why?" Hale said.

"It's not my fault you're the dumbass little piggy who made his house out of straw and I'm the smart one who used bricks," Dove replied.

Hale quit firing and rolled over onto his back. He raised a fist and then extended his middle finger at Dove. "Fuck you Sergeant. Fuck you rough and wild."

Both men laughed.

A few more minutes of fighting passed before Dove said, "When's that tank retriever suppose to get here?"

"How am I supposed to know?" Hale replied.

Dove looked up at the sky above them and saw a Cobra make another attack run with its guns on an Iraqi position he couldn't see from his current position.

"Hopefully, it won't take them too long to get here," he said. "I might be enjoying the fight, but I don't like being exposed like this. Sooner or later, some of these bullets are going to get lucky and start hitting us. The sooner those boys get here, the sooner we can load back up in the tracks and have some protection."

Hale ejected an empty magazine from his weapon, allowed it to fall to the ground, pulled another out of his gear, tapped it on his helmet to make sure the rounds inside were properly seated and loaded it. With his weapon reloaded, he went back to fighting.

"Yeah, but look at the bright side. Without those flyby boys in the Cobras, we'd really be up shit creek without a paddle," Hale said after a few minutes. "And no, that wasn't meant as a play on where we are right now either."

Dove chuckled.

An hour later the tank retriever, an AAV that had been heavily modified to tow vehicles, including heavy vehicles like tanks, arrived, escorted by two Humvees armed with fifty-caliber machine guns and one with a Mark-19 grenade launcher. The Marines around the sewage wanted to cheer for joy at the sight, but they couldn't because they were under too heavy fire.

Hale did take a quick second to glance at the tank retriever and saw the track's commander walk up to Lieutenant Owings, exchange a few words while the two men pointed at the mired vehicles. The track commander returned to his tank retriever and the vehicle started repositioning itself to get to work. The Humvees that had escorted it took up positions to defend the small section of ground the Marines were holding.

CHAPTER TWENTY-FIVE

Queen's squad took up position outside another house and prepared to take its door. So far, they had taken the doors of ten houses.

Everyone was positioned the same way they had been for each house they had cleared. Queen was right outside the door with his SAW gunners right behind him, ready to take out anyone on the other side who presented an imminent threat, and the rest of his squad was stacked up on either side of the door, ready to charge in.

The lance corporal gave the door a solid kick and it flew open with little resistance. Having learned his lesson from the close call he had had at the first house his squad had taken, Queen quickly swung around to the side of the door so he was clear of any firing that took place after he kicked the door open.

This time, no one fired on them as soon as the door swung open and the squad quickly filled the first room. In that room they found a family of five huddled in the corner. Both parents appeared to be in their late twenties to early thirties and the children appeared to range in age from around five years old to around eleven. The children were between their parents who used their arms and bodies to protectively shield their offspring the best they could manage.

Immediately upon the Marines' abrupt arrival, the wife screamed and then began to cry, as did all three of their children. The husband on the other hand just pointed up the stairs and said something in Arabic, a language none of the Marines understood.

"Shhhhhhh," Queen said with a finger up to his lips. "It's ok, we won't hurt you. We're here to help you. Now, please be quite."

He hadn't expected any of them to understand him and none of them did. Instead of calming down, as Queen had hoped they would, the family just kept making the same amount of noise they had been since the Marines' arrival in their home, quite possibly ever since the battle had begun.

Looking around, Queen pointed at a private. "Dole, stay here with them. Keep them in this room so they don't get hurt," he said.

"Gotcha," Private Roger Dole replied.

"Everyone else, break down by fire team and secure this floor," he ordered.

Within seconds the word, "Clear," began ringing out as each room on the first floor of the house was searched and eliminated as a possible threat. Deep down, they really hadn't expected to face any aggressors on the first floor since the gunfire they had witnessed coming from this house had been from a second floor window.

But then, the time they had spent in El Jasiph so far had taught them all to expect the unexpected. Out on the street, danger could lurk behind every courtyard wall, in every doorway or window or down every alley. They had already learned that in a house, danger could exist in any room, in any closet, behind any door, under any bed and on any roof. It would have been the height of foolishness, quite possibly the last mistake any of them would ever have the chance to make, if they automatically assumed a room was safe just because they had not seen anyone firing at the rest of the company from it.

The squad regrouped at the base of the stairs. With the exception of the actual entry itself, going up the stairs to the next floor was the most dangerous part in securing a house. If someone up there started to fire at the Marines, they could take cover from the Marines' return fire, but the Marines didn't enjoy that luxury.

To lessen that danger, the rest of the squad covered Queen as he crept half way up the stairs as quietly as he could. Once he was in position, Queen pulled out a flash bang grenade, which was a non-lethal weapon that disoriented those around it with a loud bang and a flash of very bright light when it exploded. It was still possible that someone could be injured by the shrapnel caused by the explosion, but the explosive force was less than that of a standard hand grenade. He pulled the pin on the flash bang and tossed it onto the second level.

He didn't know if there were any Iraqis around it or not, but if there were, it would allow him and his men to get to the second floor without getting shot. It would allow them to take anyone who was dazed by it prisoner, which was their objective secondary to eliminating threats to the rest of the company. The third advantage of using a less than lethal device was if there were any noncombatants around when the grenade went off, they wouldn't be killed, just briefly disorientated, and possibly injured from the shrapnel that flew through the air.

Once the grenade blew, the Marines didn't waste any time getting to the top of the stairs. If there was anyone up there, they were just momentarily disoriented, not permanently disabled, so their window of safety was relatively small.

They didn't find anyone at the top of the stairs and quickly began to spread out in fire teams to thoroughly search the second floor. Within seconds the sound of an Iraqi AK-47 being fired barked out, which was almost immediately answered by an American M-16 and the thud of a body hitting the floor.

Queen turned his head to see what had happened and saw one of the Marines in his squad with his right hand over his left arm just below the shoulder. Even from the distance that separated them, Queen could tell that the man was bleeding pretty heavily.

"How bad ya hit Will!" he yelled.

"Better than the camel fucker that tagged me!" Private First Class Will Gavan answered.

"Answer me, shithead!" Queen replied.

"Just winged. I've had worse playing football in high school. It's probably not even bad enough to get me a Purple Heart."

The squad leader wasn't sure if he could trust Gavan's assessment of his own wound or not. "Dole! Check his arm out!" Queen ordered.

Private Roger Dole, the closest Marine to Gavan, hurried over to him and examined the wound with his untrained eye.

"Don't look too bad! Don't think the bullet is in him, looks more like it nicked him pretty good as it flew by. Bleedin' kinda good though, should probably have Doc O'Neill take a look at it," Dole reported.

Gavan's head turned to Queen. His eyes were begging him not to send him out of the building.

"You heard him Will, get out of here. Go find Doc," Queen said.

"Come on, I'm fine. Like I said, not even enough for a Purple Heart," Gavan replied with a hint of begging in his voice.

"I've got too much to worry about in here already, the last thing I need is for you to fall out because you lost too much blood. Now, get your ass out of here," Queen said.

Gavan just looked at him with his pleading eyes. He really did not want to be forced out of the war this early. What would he tell the people back home?

"You heard me! Beat it!" Queen barked with the ferocity of the drill instructors who had ordered Queen around during his time at the Marine Corps Recruit Depot San Diego. Dejected, Gavan slung his M-16 over his right shoulder so he could continue to use his right hand to fight loosing blood from his left arm, left the building in search of Petty Officer Third Class Michael O'Neill, Alpha Company's corpsman. He knew it was the right call. He was injured worse than he had let on and doubted he would have been worth very much in a firefight, but he didn't want to leave. He wanted to be with his squad. Wounded or not, he didn't feel right leaving them.

As he left, the rest of the squad finished clearing the second floor without any further incident. At the stairway to the roof, the Marines experienced the same fear they experienced before taking any door or going up any stairwell, the fear of not knowing what was waiting for them. It was fate's little grab bag, but unlike the grab bags their parents used to buy them as children, if they didn't like what they found, it could end up costing them a lot more than a dollar.

Everyone took up their now well-accustomed positions and, as soon as everyone was in place, Queen began his creep half way up the stairs. In his hand this time was an explosive grenade, since the sun would mute the effect of a flash bang. Using an explosive grenade on the roof didn't bother him since they had found what he assumed to be the entire family on the first floor and the gunman they had seen in this house had been on the roof.

Once he was where he needed to be, Queen pulled the pin on his live grenade and tossed it onto the roof. In the tight confines of the staircase, the explosion was almost deafening and chunks of mud brick rained down on him.

The Marines didn't waste any time. As soon as the grenade exploded, they charged up the stairwell and began to fan out on the roof. They didn't find anyone up there.

"Where the hell is he?" one of them asked.

"Maybe he changed positions, like snipers do. Maybe he was the one who tagged Gavan," Private First Class Karn said.

Looking around the roof, Queen nodded his head. "That had to be him. No one else here but the family. Let's clear outta here."

Gunnery Sergeant John Ganton, Corporal Troy Jandris and Private First Class Jason Campbell secured an alley four houses down the street from the house Queen's squad had just cleared. Second Lieutenant Frank Karsen, their Platoon Commander, had put them there to deny the Iraqi forces the ability to use it for hit and run raids on Alpha Company. Several of the alleyways were similarly manned.

An Iraqi soldier came out of an adjoining alley and was stunned to see three U.S. Marines in his way. He brought his AK-47 up to his shoulder and was able to fire off several panicked and unaimed rounds in the direction of the Marines.

All three Marines returned fire and the Iraqi soldier fell dead, a mangled and bloody mess. As he was being hit, one of his bullets ricocheted off of a cinder block wall and hit Corporal Troy Jandris in the right side of his forehead. The bullet came to a stop in Jandris's brain. The thud of his hitting the ground was drowned by the noise of both Ganton and Campbell firing their weapons.

Without looking back towards either of the men with him, Gunny Ganton cautiously moved towards the Iraqi's body with his nine-millimeter Beretta ready to be fired at the slightest movement. The first thing he did when he made it to the corpse was to make sure there wasn't anyone else in the alley the Iraqi had come from, waiting to ambush the first American to show up. Once he was positive he was reasonably safe, for the moment anyway, Ganton bumped his toe into the dead Iraqi several times, just to make sure he was dead.

Ganton turned around and saw Jandris lying face down in the Iraqi sand. "Christ! Jandris!" he yelled and ran towards the fallen man.

Campbell had been too busy covering Ganton while he made sure the Iraqi soldier was dead and hadn't even though about Jandris, who was behind him, until he saw Ganton's reaction. He turned and saw Jandris, lying on the unpaved street with blood pouring out of his head. "Holy Mother Mary," were the only words his mind could think to form.

Immediately after arriving at Jandris's side, Ganton noticed his blood wasn't simply pouring out of him but that instead it was being actively pumped out of the wound in his head.

He's... he's still alive, Ganton thought and quickly reached for Jandris's right wrist to feel for a pulse.

It took him a few seconds to find the wounded man's pulse, but he found one. It was weak and thready but it was there. He was alive, maybe just barely, but he was still alive.

"Campbell! Go get Doc O'Neill! He's still alive!" Ganton ordered.

The twenty-year-old private had been standing there, watching stupidly as his Gunnery Sergeant checked out Jandris. He had thought that he had seen his first friend die. Cambell's spirits rose like a helium balloon when he heard Ganton's pronouncement that Jandris was still alive. He didn't need any more incentive. Campbell wanted his friend to live and took off running to find the corpsman.

Wishing he had an M-16 like almost everyone else, Ganton brought his pistol back up and went back to work covering the alley. He knew there wasn't anything he could do for Jandris. The Marine would live or die and it was out of his hands, but he did know the young man didn't have a chance if the Iraqi soldiers were to take control of the alley.

CHAPTER TWENTY-SIX

Even though they had made it through the gauntlet of El Jasiph, Charlie Company was still under heavy enemy fire and it felt to Captain Brian Aber that his company was taking an extraordinarily high number of causalities. It seemed that everywhere he looked, he saw some of his Marines on the ground, either screaming and writhing in pain or lying still in the silence of death.

This plan sure fell apart quickly, he thought as he observed his company's position. *Alpha and Bravo were supposed to have come in from the flanks and we were all supposed to be here together right now, raining death down on Haji. Where are they? Things weren't supposed to go this badly.*

He didn't know the details, but Aber knew the other two companies were in trouble too. As they fought their way through El Jasiph, Aber had been more concerned about what was being said on the Company TAC since his company was his responsibility. Every now and then he had listened in on the Battalion TAC, just to see what was happening to Alpha and Bravo Companies. They were his brother Marines after all. How could he not care what was happening to them? Not to mention that their mission was to secure El Jasiph for the battalions that would follow them during this war. That meant if any of the companies were decimated, the rest of them would have to re-fight that ground so those who came next could get through the city without having this happen to them.

From his first day as a plebe at Annapolis, Aber had dreamed of leading men into battle. That's what Marine officers were trained for. However, like most of the men under his command, he had not experienced battle until El Jasiph and didn't really have any idea what it would be like. He was scared, but he had a job to do and he refused to allow his fear to rule him. Fear was something that was to be mastered and brought to heel; it wasn't something that an officer in the United States Marine Corps allowed to master him.

Aber knew he had men wounded, many of whom were severely wounded. He also knew he had men who had started the day alive, but who were now dead. That was something he knew would happen, something they had prepared him for in the academy, but it was also something he had not really understood until the first report of a fatality had reached him.

Charlie Company was taking a beating. That much would have been clear to anyone with even just one working eye who looked around and saw the company's corpsmen running through the incoming Iraqi bullets, from wounded Marine to wounded Marine, doing the best he could to keep each of Aber's wounded men alive. His company was against the ropes, but everyone, from his Platoon Commanders to the youngest privates just out of high school, were maintaining their combat discipline and they were fighting exactly as they had been trained to do. They were doing the United States of America proud, they were doing the United States Marine Corps proud and they were doing themselves proud. For that, his pride in each and every one of his men was immeasurable, even if he didn't know all of them by name.

Petty Officer Third Class Craig Trevino ran around the battlefield, carrying his large, backpack-like medical kit. The battle had turned so bad on them, and there were so many Marines wounded now that he had stopped carrying his medical kit on his back, it was just too much work to get it on and off in between patients. Instead, he now carried it in his hands as he ran from wounded Marine to wounded Marine. Despite how bad things had gotten, he still had not removed his nine-millimeter from its holster.

Trevino came to a young Marine with a left arm that had been shredded so badly it no longer resembled an arm. He placed a tourniquet above the wound and pulled a pre-filled syringe of morphine from his medical kit. He uncapped the needle and plunged it into the Marine's leg without bothering to remove the Marine's sand-covered pants. His objective was to do what he could to relieve the man's discomfort and move on to the next wounded man as fast as he could. Once the syringe was empty, Trevino allowed it to fall to the ground and pulled a black marker out of his pocket. He wrote M-0830 on the man's forehead to remind himself he had given him morphine at 8:30 A.M. If there were any possibility of saving the marine's arm, he also would have written T-0830 on the Marine's forehead as a reminder of when he would have to release the tourniquet so blood could flow back into the wounded limb. It was hoped that occasionally allowing blood flow back into the wounded limb would allow the surgeons at the battalion aide station or a MASH to save the limb. However, even a casual glance at the man's arm was enough to tell even an untrained eye that the limb was beyond saving, so he didn't bother to make that note. It was better to keep as much of the Marine's blood in him as he could.

The corpsman moved on to the next wounded man and saw he had a large piece of shrapnel protruding from his chest, too close to his heart and possibly even through a lung. His flesh was already deathly pale and each breath that came out of his mouth carried blood with it. There was nothing Trevino could do for him. This Marine would be going home in a flag-draped coffin. Trevino gave the man a dose of morphine to make his last moments as comfortable as possible and drew a large X on his forehead.

"What the fuck does that X mean?" a Marine next to the dying man asked.

Trevino looked at him. "It reminds me that I've already tried to save him and there's nothing I can do."

"What the... fuck no! You ain't givin' up on Max like that!" the Marine said. "He's getting married when he gets home. He can't die!"

All Trevino could do was shrug sadly. "I'm sorry, but I'm not God. There's only so much I can do, then it's out of my hands."

The Marine took a second look at his friend, who was knocking on death's door. When that second passed, the dying man's buddy turned away from his friend and Trevino and went back to tending to the wounded. There would be time for mourning lost friends later.

Trevino picked up his pack and ran on to the next wounded Marine.

CHAPTER TWENTY-SEVEN

The hunting from the roof where the snipers had made their perch had dried up, the Iraqis had learned the snipers were there and started being careful not to make too good a target of themselves. So the snipers assigned to Alpha Company broke up into two-man teams and found new places to conceal themselves and take out Iraqi soldiers.

Corporal Rob Beverly and Sergeant Henry Quartermaine found a place behind a courtyard wall. Beverly had the barrel of his sniper rifle sticking through a hole the size of a Coke can, which had been caused by a fifty-millimeter round punching through the wall.

He scanned the street and nearby buildings for good targets, Iraqis who he doubted the rest of the company would be able to take out and who had the potential of killing a lot of Marines. There were plenty of targets to choose from, which wasn't a problem, but a sniper's job was not to simply kill any Iraqi he saw. His job was to take out the most immediate threats to the company, the threats those pinned down on the streets simply didn't have the time to locate, isolate and target as individuals.

As he scanned the city in front of him, Beverly's attention fell on a young boy, who appeared to only be seven or eight years old, who had an AK-47 in his hands. Like almost every other Marine in El Jasiph, he had seen the movie Black Hawk down and knew children had killed American soldiers in the Somali city alongside their adult family members. There were so many similarities between that movie and El Jasiph, it couldn't help but weigh on the minds of the Marines there, so Beverly allowed his scope's crosshairs to come to rest on the child in the sniper's triangle and rested his finger on his rifle's trigger.

Something's not right, Beverly thought as he watched the little boy through his scope, ready to fire. *He's not holding it like a combatant. He looks scared, he doesn't know what is happening around him.*

Beverly continued to observe the boy, trying to determine if he was a danger to the Alpha Marines, or if the boy was the one in danger of being mistakenly identified by someone as a hostile and killed.

Figuring out what was going on with the boy took a lot more time than Beverly cared to spend on any one particular target, but he wanted to get this right. The last thing he wanted to do was to accidentally kill an innocent civilian, especially a child. At the same time, however, he didn't want to hurriedly classify the boy as a non-threat and move on to someone else, only to have that same child turn around and start killing Marines in a few minutes.

Several minutes, which felt like hours to Beverly, went by and the boy didn't once raise the weapon in his hands to his shoulder and didn't do anything else to give the sniper the idea that this kid was unquestionably a threat. As a matter of fact, he became more and more convinced with each passing second that the boy was nothing more than an innocent child who hadn't asked for the war to come to the city he lived in. He was lost, confused and scared. He might not even be sure what it was he was carrying. The boy certainly didn't look as if he knew how to use it.

Come on, the sniper thought, *drop that weapon and go home. Get lost. Beat it. Please.*

Everything about the boy further served to confirm Beverly's suspicion that the kid was not a combatant, that he was just an innocent in the wrong place at the wrong time, which was a mistake that might cost him his life at such a tender age. He wondered about the boy's parents. How could they let their young son wonder the streets alone while a battle was being fought in the city? Then it occurred to him the boy's parents might have already been killed and that he might have witnessed their deaths.

Come on kid! Get out of here! Beverly thought as he continued to watch the young boy. Bullets were flying everywhere, so the sniper knew each and every second that passed brought the kid another second closer to death.

Beverly dropped his crosshairs to the sand at the boy's feet and squeezed the trigger. His bullet raced from the weapon's barrel, cut through the air, crossed through the battle and struck the sand, making the little boy jump.

The bullet hitting near him had obviously rattled the boy, but he didn't run. Instead, he just stood there, stupefied without any real comprehension of what was going on around him. Beverly chambered another round and once again targeted the ground right in front of the boy. He pulled his trigger.

This time the bullet had its intended effect. The second impact had been enough to shake the boy from the daze he had been in. He screamed, dropped the AK-47 he had been clutching and took off running away from the battle.

Beverly smiled and hoped the kid would live to see tomorrow.

brought . . . across . . . through the blind . . . see . . . of the son . . .
make the little boy . . . angry . . .

The still burning note him loudly . . . every turn of his leg . . . but his
interruption . . . He just stood there, still . . . his will, but as his . . .
disturbed . . . her mind away . . . to tie . . . back . . . pull . . . in
. . . back here. Trampled his fingers . . .

The jumping and . . . let . . . reason and effect . . . so obviously had . . .
by . . . with gentle but thorough . . . to he had been in the
course . . . during the A . . . Both I have absolutely unlocked
. . . him away down the road . . .

. . . book . . . from the kid. And he . . . to see I am now . . .

CHAPTER TWENTY-EIGHT

Four of Charlie Company's M249 SAW operators had taken up positions in the sand next to each other in the belief that all their heavy machine guns firing into the same area would do the company more good than if they were all separated from each other.

"Shit!" Lance Corporal Tim Newtson, one of the SAW gunners, yelled. "Cliff! Hey Cliff! I'm dry man! Get me more ammo!"

"Don't got no more!" Private Cliff Orchard replied. Orchard, Newtson's "A" gunner, knelt behind him as he fired his M-16.

"What!" Newtson asked, clearly displeased.

"I'm tapped! No more ammo for you!" Orchard answered as he fired his weapon.

Since he didn't have any more bullets to fire at the Iraqi soldiers, Newtson turned around to look at his "A" gunner. Between the sand that caked his face and the fury that burnt in his eyes, he reminded the younger private of a demon from a horror movie.

"Well, get your ass back to the track and get some more! Damn man, how dumb are you! Your job is to keep me firing! If this were an exercise and not the real fucking deal, I'd be kicking your mother fucking ass right now!" Newtson barked.

Orchard hated being talked to like that and he really hated anyone telling him what to do. It was something he rarely tolerated, especially from a lance corporal who wasn't too much farther up the totem pole than he was. Orchard knelt there and stared at Newtson for a brief second, as if he were deciding if he should start the fight himself. It took less than a second though for the expression on Newtson's face and the hateful glare in his eyes to convince Orchard that at that moment it time, it was quite possible he was in more danger from his fellow Marine than he was from any Iraqi.

Deciding it was the wrong time and place for a pissing match, but making a mental vow to settle the score later, Orchard slung his M-16 over his shoulder and took off towards one of the tracks. The other three "A" gunners ran with him, deciding they should get more ammunition for their SAW operators before they ran out.

Seconds later, the four "A" gunners entered the track and they each quickly grabbed a box of the belt-fed ammunition the SAWs required. Once they had what they'd come for, all four of them left the track and took off running back towards their SAW operators.

With everything going on around them, not to mention all of the noise of battle, none of the Marines saw the mortar round high above them as it began its downward trajectory. The mortar struck the ground between them, sending an eruption of sand into the air as it exploded. The explosion threw all four "A" gunners through the air, as if they were nothing more than G. I. Joe action figures unfortunate enough to be caught in the middle of a spoiled child's temper tantrum.

In the aftermath of the explosion, Private Cliff Orchard lay on the ground, his eyes staring up lifelessly at the blue sky above him. His torso had been torn wide open, as if he were an unfortunate victim in some particularly gory horror movie.

As if he wasn't already overworked enough, Petty Officer Third Class Trevino saw the mortar strike the ground and sprinted to the four "A" gunners. First he saw Orchard's corpse and could tell from just a quick glance it was already too late for him so Trevino ran to another "A" Gunner, who appeared to be intact.

He didn't stay long though. He rolled the Marine, who was face down in the Iraqi sand, over and just one look at his eyes was enough to tell Trevino it was too late for him too.

Trevino didn't linger over the dead man. Instead he ran to the next "A" gunner. Despite his injuries, Trevino was excited to see one of them alive, but he would have to act fast to keep him among the living.

The first thing Trevino did was reach into his medical kit and pull out two tourniquets, which he didn't waste any time placing above the bloody stumps where the Marine's legs had been just seconds before.

"Oh, shit, Doc! I'm dying ain't I," the Marine yelled, panic and pain both heavy in his voice.

"No you're not. You're not hurt too bad at all," Trevino lied as he put the tourniquets in place to stop the man from losing any more blood than he already had.

"Don't bullshit me, Doc!" the Marine yelled.

As soon as the tourniquets were secured in place, Trevino pulled out another pre-filled syringe of morphine.

"Not bullshitting you Marine," Trevino said as he injected the narcotic into the injured man. "You've served our country proudly and it won't be too long at all until you're back home and using your Purple Heart to convince the pretty little things at the dance club to let you into their pants."

The morphine started working. Free of the inhuman pain he had been in, the Marine was able to visibly relax. He was higher than a kite and wherever he was, his mind had taken him far away from the battle of El Jasiph. In a way, Trevino envied him that.

With the necessary work done, Trevino pulled the black marker from his pocket and noted the time he had injected the Marine with morphine. He took satisfaction in knowing this one was likely to make it home, granted without the legs he had come to Iraq with, but he wouldn't be one of the young men giving their lives in this desert. He would once again get to see his friends and family, kiss his girlfriend, maybe have a family of his own and continue on with his life. With luck, the Marine wouldn't die until he was an old, old man.

There wasn't time for Trevino to stay in one place and enjoy that feeling of satisfaction. Charlie Company was being hit hard and there were wounded and dying men all around him so he had to keep moving. He ran to the fourth "A" gunner, who was also clearly deceased.

Having done as much as he could for the four "A" gunners, Trevino picked up his medical kit and ran to the next closest Marine who was screaming in pain.

CHAPTER TWENTY-NINE

Petty Officer Third Class Michael O'Neill stood, impatiently listening as Alpha Company's First Sergeant, Hector Urness, called for a medevac from inside the company's command track. His heart sank as he heard the battalion aide station, which was on the other side of the Euphrates River, reply that their location was too hot for them to risk landing a helicopter to pick up wounded.

"I'll take him in a Hummer then," O'Neill said as he turned to leave the track.

"No you won't,' Captain Dave Callen said, bring the Corpsman to a halt.

The medic turned around to look at Alpha Company's Commanding Officer.

"But, sir, he's been shot in the head, he is in critical condition. If he doesn't get to a doctor soon, he's not gonna make it. We need to get him out of here and to the battalion aide station immediately," O'Neill said.

"No, and that's final," Callen said.

"But, sir... " O'Neill continued.

Callen brought his protest to a stop with a wave of his hand. "And what happens if I let you take him? Huh? I let you take a Humvee and all of a sudden I'm short the fifty that's on top of it and my corpsman. How many men will die then?"

"Sir, I can... " O'Neill tried again.

"The answer is no, Petty Officer," Callen said, cutting him off again. "You will stay here and do what you can for those who get hit. I hate the idea of losing any of my men, but I won't risk losing any more while you're gone so that you can run off and play ambulance driver. You stay, and if Corporal Jandris dies, then he dies. A sad fact about war is that Marines die fighting. Your job is to do what you can to make sure that fewer Marines die in this fight, and it looks like my job is to keep you here doing yours."

Realizing he had lost the argument, O'Neill left the AAV and went back to work tending to the wounded.

Out on the street, Sergeant Tom Daley ran from position to position, doing his best to avoid being hit by any Iraqi bullets himself. He came to one of the CAAT Humvees and looked up at Private First Class Brad Barnebee, who had been pumping fifty caliber bullets into the doors, windows and walls of the buildings that had Iraqi soldiers inside of them non-stop since Alpha Company had first come under fire.

"Private!" Daley yelled up at Barnebee, to be heard over the noise of weapons being fired around them.

Barnebee didn't want to acknowledge the Sergeant. He didn't want to stop firing his weapon because he was in an exposed position and really didn't want to get hit, but at the same time, he knew better than to ignore a sergeant. He dropped down into the Humvee so he wasn't sitting on top of it while talking to the sergeant and went to the door.

"What can I do for ya, Sarge?" he asked.

"You been drinking?" Daley asked.

Confused by the question, Barnebee replied in a mock-slurred voice, "Why, yes, Sergeant, I certainly have been. Just made myself a Bloody Mary if you'd like one."

"Water, smart ass!" Daley said, annoyed by Barnebee's joke. "Have you been drinking water?"

It dawned on Barnebee that he hadn't had anything to drink since about an hour before the battle started.

"No, I haven't," he answered honestly.

"Then take a minute and drink some. Drop down into the Hummer every so often and keep yourself hydrated. We don't want you going down because the heat got to you," Daley said.

Barnebee hadn't noticed it before, but his mouth and throat felt like sandpaper.

"Yes, Sergeant, I will. Thanks for reminding me," the private first class said.

"No problem," Daley said and took off running to the next nearest position.

Just as he made it to the next Marine, a lance corporal who had taken cover next to a courtyard wall, a Cobra helicopter came in from above and fired three Hellfire missiles into the surrounding buildings.

Rounds from the tanks had brought buildings down into heaps of rubble, but the Hellfire missiles from the Cobra blew the houses apart entirely, causing rubble of various size to rain down on the Marines. Some of the chunks of cinder block and mud brick were large and dangerous, so, instinctively, many of the Marines covered their heads as the rubble fell on them.

"This shit's real ain't it," the lance corporal said.

"Don't get no realer than this," Daley replied as he watched the Cobra fly past them and turn around for another run.

This time, instead of firing missiles into the houses, the helicopter strafed Iraqi soldiers who were on the rooftops with its twenty-millimeter, three-barreled Gatling gun. From where he stood, Daley saw several Iraqis fall, and even saw a few fall to the street and alleys below.

"Pretty sight, ain't it," the lance corporal said, awe filling his voice.

Daley nodded his head. "You bet it is," he said as he watched the Cobra fly off to another area of the city.

The helicopter's attack run had brought a stunned silence to that area of El Jasiph as almost every weapon fell silent. The effect was short lived though and within seconds, the deafening noise of war returned.

"You been drinking water?" Daley asked, returning his attention to the lance corporal.

"You bet I have."

"Good boy. Keep it up," Daley said and started running to the next closest Marine.

CHAPTER THIRTY

With his face in the sand, Charlie Company's Private Andy Irvin felt sand and small chunks of rock fall onto his back.

That mortar came way too damn close, he thought as he lifted his face out of the sand berm, pointed his M-16 back towards El Jasiph and started firing.

Charlie Company had been under heavy mortar attack pretty much ever since they escaped the urban maze of El Jasiph. For the most part, the Iraqi mortar shells had been way off target and fell harmlessly in the vast, seemingly endless, desert around them. Every now and then one would come close to hitting their target, just like that last one had come too close for comfort to landing on top of the position he and the rest of Charlie Company's Three Stooges were fighting from.

Even though most of the enemy mortar rounds were landing far from their mark, the Iraqi soldiers firing them apparently knew how to walk them towards the target because each shell that came in landed closer to what they were trying to hit than the one before it had. One had hit a track, which burned brightly thanks to all of the ammunition that had been stored inside it cooking off.

Irving pulled his weapon's trigger only to be greeted by the clicking sound of an empty magazine. He rolled to his back, ejected the empty magazine and replaced it with a fresh, fully loaded one.

While he was on his back and looking away from the city, Irving took a second to watch Charlie Company's mortar position. Granted, he didn't know too much about their job, but it seemed to him that they were grouped awfully close to each other. In training the mortar positions had always been spread out more than what he saw.

He watched as an Iraqi mortar shell landed about sixty feet in front of the Marine mortar position. A staff sergeant stood and looked through binoculars to see where the last shell from his position landed and shouted targeting corrections to the Marine who was actually firing the rounds. Irving didn't envy those men at all. He was laying on a sand berm, which ate up the vast majority of the bullets that were fired in his direction, but they were out in the open, completely exposed to enemy fire as they fired mortar after mortar into El Jasiph.

"What're you doin' asshole!" Irving heard Burt Farrow yell at him. "Get back into the fight!"

With a fully loaded magazine in his weapon, Irving rolled back to his stomach and resumed firing at the Iraqi soldiers that were firing at them. As he fired, Irving saw one of the American mortars fall on the roof of a house where some of the Iraqi mortars had been fired from.

"Incoming!" he heard someone yell.

Instinctively, Irving, and the rest of the Marines around him, quit firing and put their faces in the sand to protect their faces and their throats from any shrapnel that might fly around them if the mortar were to land anywhere near their position. He heard the shell hit the sand and explode and then felt sand and rock pelting him as the most recent sand geyser fell back to the ground.

As each Marine realized that they hadn't been ripped to shreds by shrapnel from the exploding mortar shell, they picked their faces up out of the sand and resumed firing at the Iraqi soldiers in El Jasiph.

Irvin heard Manny Alverez yell, "Hey, spic! You enjoyin' your visit to Camel Land Family Fun Park?"

"Shit, honkey, I think I'm going to need to visit a proctologist to get all of this sand out of me when I get home!" he heard Farrow answer.

He smirked as Alverez and Farrow laughed.

"You two are all sorts of fucked up! You know that don't you?" Irvin yelled to be heard over the noise of battle.

"Sure do, sweet tits," Alverez answered. "And we wouldn't have it any other way."

The comment made Irvin chuckle. He loved Farrow and Alverez, they were like the brothers he never had and deep down, if any of them were to die in this battle, he would have wanted it to be him because he wasn't sure that he could have handled the death of either of them.

A few more minutes passed without conversation as each of the Marines fired their weapons and changed out magazines when it was necessary. Irvin saw another American mortar fall on the roof of an Iraqi building in El Jasiph. It didn't cause the building to collapse, but he didn't have any doubt that it's roof was gone, and that it more than likely crushed the Iraqi soldier on the second floor who had been firing a heavy machine gun of some sort at them. That Iraqi's death seemed to be evidenced by the fact that there wasn't any more hostile fire coming at them from that building.

Once again Irvin heard someone yell, "Incoming!" and he buried his face in the sand without hesitation. After the now familiar pelting of sand and small rock passed, he raised his head back up and started firing.

His magazine ran out of bullets again so Irvin rolled to his back to change them out. It was a task that he didn't need to see what he was doing to accomplish so he glanced up at Charlie Company's mortar position. His eyes grew wide with horror when he saw an Iraqi mortar coming down right into the middle of their position. There wasn't anything he could do to warn them, they were too far away and the noise of battle around them was too deafening to yell a warning. The sound of his voice would simply never make it.

The poor mortar Marines didn't even see the Iraqi shell coming in until it was too late. The enemy mortar round hit the ground and blew up. The force of the explosion sent sand, rock, bodies, mortar tubes and mortar rounds up into the air, all of which landed ungracefully on the desert beneath them.

"Oh shit!" Irvin yelled without realizing that he had done it.

Farrow turned his head to see if his friend had been hit and followed his gaze when he saw the terrified expression on Irvin's face. He quickly turned his attention to Alverez and shook him.

"Come on, honkey!" Farrow said, got to his feet and took off sprinting towards the wounded Marines. Irvin and Alverez were quickly behind him.

By this time, Petty Officer Third Class Trevino had established a casualty collection point on the safest side of one of the tracks—not that any place was truly safe for any of them. The horror of war had already hit home for each of them. The Battle of El Jasiph had been going on long enough for reality to smack each of them in the face nice and hard, so the sight of mangled and bloodied bodies didn't faze any of them as it would have just twenty-four short hours earlier. They ignored the dead as they searched for those who were still alive so that they could take them to the causality collection point until a medevac out could be arranged for the wounded.

Irvin saw that staff sergeant he had watched earlier. The man had been disemboweled by a flying piece of shrapnel and stared up at the blue sky above them with eyes that no longer contained life.

The next Marine he saw was so badly wounded that Irvin made the instant decision that he was dead. The man was missing half of his face, had a nasty gash to his chest and both of his legs were bent at impossible angles, so he couldn't possibly still be among the living. Then Irvin saw the Marine's right hand move and realized he was still alive—probably at death's door, but still alive—and picked him up in a fireman's carry.

With the critically wounded man over his shoulders, Irvin ran to the casualty collection point as fast as he could with the added weight he was carrying. Farrow was about twenty feet ahead of them and the speedy little devil Alverez was well ahead of both of them.

Half way to the casualty collection point, Irvin saw a part of Farrow's left shoulder flip up like the tab of a soda can and fall. Digging down as deep as he could, Irvin put every bit of speed he could into his legs to get to his friend.

Once there, Irvin laid the man he was carrying down on the sand as gently as he could and rolled Farrow, who was lying on his right side, onto his back. Farrow was screaming in pain and Irvin could tell that pain was the only thing that was on his friend's mind.

"You're alive, thank God you're alive," Irvin said without realizing that he had said it. It was an honest expression of thanks to the man upstairs, not an attempt to start conversation with his wounded friend.

Indecision pulled at Irvin. He wanted to get Farrow help right away, but knew the Marine he had started out carrying was worse off than his friend was by far. After a brief moment of hesitation, Irvin told Farrow, "I'll be right back for you," even though he knew his friend was too wrapped up in his own agony to have heard him, put the man that he had been carrying back over his shoulders and took off running again.

As soon as he reached the casualty collection point, Irvin lowered the wounded Marine to the ground as gently as he could and saw that Trevino had made Alverez stop, sit down and drink some of the warm water in his canteen. Ordinarily, several jokes at Alverez's expense would have sprung to mind, but not then and there. Irvin didn't even say a word to him. Instead, he took off back towards where Farrow still lay on the ground, bleeding heavily. It was almost like Irvin was running with horse blinders on, he didn't even notice the little plums of sand that were being kicked up around his feet by Iraqi bullets.

Irvin picked Farrow up in a fireman's carry and ran back to the causality collection point. There, he laid Farrow on the ground next to the first man he had carried in, who now had a black X on his forehead, and ran back out into the fray to recover the Marine Farrow had been carrying.

An Iraqi mortar landed fifteen feet in front of Irvin as he ran to the wounded man. The force of the explosion picked Irvin up off of the ground and threw him backward ten feet. He landed hard on his back and all of the air was forced out of his lungs.

Irvin's chest burned from the lack of oxygen, but his lungs quickly refilled themselves and he examined himself. As soon as he realized he was uninjured, Irvin stood back up and continued running to the wounded Marine.

CHAPTER THIRTY-ONE

Sergeant Brian Martin, Bravo Company's forward air controller, had been in contact with the Cobra helicopter that had been providing air cover for the vehicles that were mired in sewage while the tank retriever worked to free them from the muck. They had been under constant enemy fire, but Martin believed wholeheartedly that it would have been much worse if the attack helicopters hadn't been in the sky above them. The Iraqis didn't seem to like the Cobra's twenty millimeter Gatling gun, Hellfire missiles, Sidewinder missiles and rocket pods very well.

The helicopter came in for another pass and laid down strafing fire along the city streets with its gun, sending the Iraqi soldiers on them scurrying for cover.

"Vampire to Tomcat Eight," Martin heard the pilot of the Cobra helicopter above them say.

"Tomcat Eight, go ahead Vampire," he replied.

"I'm Winchester. I've got to return to base and rearm," the pilot said.

Winchester meant that the helicopter was completely out of ammunition. The words caused Martin's heart to sink. He didn't have any doubt that the incoming Iraqi fire would become much worse than anything they had seen yet as soon as the Cobra disappeared and the Bravo Marines who had been left behind to guard the vehicles had precious little cover as it was. The last thing any of them needed was for Vampire to disappear.

A few seconds passed before Martin replied, "Roger, Vampire, you're Winchester. Could I get you to pretend not to be?"

"Repeat your last Tomcat Eight," the pilot said, confusion heavy in his voice.

"You scare these guys shitless," Martin said. "Things would be a lot worse down here if it weren't for you."

"Tomcat Eight, I'm Winchester. I'm like a lion without teeth or claws up here. As soon as Haji realizes that, they won't be very scared of me and I'll become an even bigger target for RPGs than I already am," the pilot replied.

"Vampire, Haji is scared of you. Every time you make a run, they duck and cover and then check their underwear once you've passed. They won't even notice you're not firing at them," Martin said, the desperation he felt was starting to show through in his voice.

Several seconds of silence passed, but Vampire didn't disappear from the sky above him. Martin could almost sense the indecision on the part of the pilot above him and he couldn't blame the man any. If their roles had been reversed, Martin wasn't so sure that he would stick around if he was out of ammunition. This battle was a dangerous enough as it was, but not having anything to defending yourself with just made it all that much more dangerous and Martin wouldn't have been able to fault the pilot if he decided to fly away.

Please, please, please say you'll stay, Martin thought. Then, even with the battle raging around him, Martin recalled thinking those same very words eight months before he shipped out to Iraq when his live in girlfriend of two years moved out.

The seconds seemed so much longer. For Martin, waiting to learn what the pilot's decision was comparable to waiting for the executioner to swing his axe. Deep down, the sergeant really didn't think that the pilot would stick around, but one could always hope. After what seemed like an eternity, the pilot's voice said, "Vampire to Tomcat Eight."

"Tomcat Eight, go ahead Vampire," Martin replied.

"I'll stick around on one condition," the pilot said.

"What's that?" Martin asked, almost afraid that the pilot would ask for something that was impossible for him to guarantee.

"That you guys take out anyone that fires an RPG at me. I don't like this situation at all, but I like the situation you guys are in down there even less," the pilot said.

Martin felt as if he had been drowning and someone had put a life vest on him that brought him back to the water's surface. He smiled. "You drive a hard bargain there Vampire but you've got a deal," he replied, his voice made it clear that he was joking. "Oh yeah, and one last thing Tomcat Eight," the pilot said.

"What's that?" Martin asked.

"I just want you to be aware of the fact that if I get shot down, I will be finding out who you are and kicking your ass."

Laughing, Martin replied, "Roger that Vampire. Thank you."

The attack helicopter swooped in from above again. Coming in hot towards the Iraqi soldiers, the Cobra looked every bit as deadly as it had been when it was armed. The difference was, now the pilot was taking a huge risk because he didn't have any way to defend himself and Martin hoped that the pilot, whoever Vampire really was, would be awarded the Congressional Medal of Honor for the risk he was taking.

Feeling better about his situation than he would have if the helicopter had left, Sergeant Brian Martin went back to firing his M-16.

CHAPTER THIRTY-TWO

Charlie Company had undoubtedly seen better days. Captain Brian Aber couldn't believe the mess his Marines were in; not that it was any of their faults. In fact, they were doing the best they could under the circumstances and he had nothing but the highest respect for each and every one of them. Iraqi mortars had destroyed two of his tracks and three of his Humvees. More of his men had been killed or wounded by Iraqi bullets and mortars than he would have ever imagined he would lose the first time he led men into war. The truth be told, deep down, Aber had never imagined real war being this horrifying and while he had known intellectually that men would die under his command, there had always been a part of him that hadn't truly believed that he would lose anyone.

To add insult to injury, the Battalion TAC was still a traffic jam of voices and there simply wasn't any cutting in on them. Aber didn't have any way to call for a medevac to get his dead and wounded out.

There were days he wished that he had joined the Marines as an enlisted man instead of going through four years at Annapolis, and this day was certainly shaping up to be one of them. There were just times that being in command wasn't all it was cracked up to be.

Aber's magazine ran dry. He ejected it from his M-16 and replaced it with a fresh one. Just as he started firing again, an idea came to mind, he knew how to get his dead, and especially his wounded, Marines to the Battalion Aid Station. Aber set off in search of his First Sergeant, Randy Raab. As confusing as things had become, he'd lost track of where his First Sergeant was.

The captain slung his weapon over his shoulder and took off running through the storm of Iraqi bullets. The first Marine he ran came to was Private Charles Schuman.

"Have you seen First Sergeant Raab?" Aber asked.

"I think he's helping Doc Trevino at the casualty collection point, sir," Schuman answered.

"Thanks," Aber said and then took off running towards where Petty Officer Third Class Craig Trevino was doing everything he could to keep the wounded alive. He hadn't been successful with all of them.

Covered in sand and a sickly mixture of dried, tacky and still wet blood, First Sergeant Raab kept his bare hands applying pressure to the right leg of a private so that he wouldn't bleed to death. He had long since exhausted the supplies in his basic combat first aid kit, which all Marines carry, and hoped that Trevino would get to him before the young man, who wasn't even out of his teens yet, bled out and died.

Raab heard Captain Aber bellow out, "Doc! With me!" before he saw his commanding officer walking towards him.

"First Sergeant," Aber said when the two men reached him. "We're going to have to send the dead and wounded out on tracks since I can't get a medevac in here. I want you to get men to help Doc load them up and find fifty volunteers to ride along in other tracks to provide security."

"Yes, sir, but is that a good idea? I mean, sending them back through El Jasiph. We didn't do too well coming through the first time," Raab replied, still applying pressure to the wounded Marine's leg.

"I really don't want to send them back through the city, but we don't have any other options. Not if we're going to get them to the battalion aid station while they're still alive," Aber answered.

"I'll get on it right away, sir," Raab said and then turned his attention to Trevino. "Can I get you to take over here, Doc? This guy is bleeding pretty bad and won't last long if I take my hands off."

The corpsman dropped to his knees next to Raab and said, "I've got it, First Sergeant."

Without another word being said, Raab disappeared into the battle to find fifty men to escort the wounded to the Battalion Aid Station.

Five AAVs had been given to the medevac convoy. The first and the third carried the dead and wounded and the second, fourth and fifth carried the Marines who had volunteered to defend those who couldn't defend themselves if they got bogged down by Iraqi fire, or had to stop for any other reason.

In the distance, Captain Aber saw an A-10 Warthog, a U.S. Air Force jet that was designed for killing enemy tanks, which looked as if it were lined up for an attack run on their position.

God, please let my imagination be getting the best of me, Aber thought. *We're already taking a bad enough pounding as it is. The last thing we need is for one of those boys to start raining death on us from above.*

Unable to pull his eyes away from it, the commanding officer of Charlie Company watched the A-10 came closer and closer to his position. A solid knot of fear developed in the pit of his stomach.

Maybe they're on their way somewhere else. Maybe they'll just over fly us and keep on moving to where they are going, he thought, trying to be optimistic even though deep down he knew exactly where that aircraft was heading. It was heading right to Charlie Company.

He knew he should have his men throwing red smoke grenades to let the pilot know that they were on the same side, but he didn't want to waste the smoke if it was just passing over them.

The advantage to traveling by air over traveling by land is that you can move much faster and before Aber knew it, the A-10 was over his position.

Thank God, he thought once the jet had passed half way over his company's position. *He is just flying by and not on an attack run.*

His relief had been premature however. The A-10 dropped one of its precision-guided bombs. It felt as if Aber's heart stopped as he watched the bomb fall through the air. It hit one of his tracks, which fortunately was unmanned at the time, its crew having been needed to help load the wounded onto the evac tracks.

The bomb hit the top of the track and it exploded with all of the ferocity of a volcano, sending deadly pieces of metal flying through the air of Charlie Company's position. Aber heard his Marines, most of which had been too busy to have even noticed the A-10's approach, yell various startled curses of surprise at the explosion.

"Pop red smoke! Pop red smoke now!" he yelled as loud as his lungs would allow him too. His order was picked up and repeated by everyone who heard him and, almost instantly, smoke grenades were being thrown and red smoke began drifting from them and up into the air.

Red smoke was a non-verbal way for United States Marines and soldiers on the ground to tell pilots, "We're on your side, stop shooting at us."

Aber watched as the A-10 finished flying over their position, turn around and start to come back towards them.

He's gotta see the smoke. He won't fire on us again, Aber thought optimistically he watched it approach.

Just as his relief had been, Aber's optimism proved to be every bit as premature. With all of the smoke drifting up into the air, there simply wasn't any way the pilot could have missed it. Maybe he thought the Marines of Charlie Company were Iraqi soldiers trying to trick him into aborting his attack run. Whatever his reasons were, the pilot ignored the smoke and sent bullets from his aircraft's rapid fire anti-tank gun into another of Charlie Company's AAVs. If tanks couldn't stand up to those rounds, the thin skinned track didn't have a chance. The A-10's bullets passed through it, like a fist through a sheet of paper, and kept on going.

Scared worse than he had ever been in his life, Captain Aber ran to the front of the medevac convoy and banged on the driver's side door of the lead track.

"Get out of here goddamnit!" he yelled. "Get the fuck out of here now!"

The driver didn't need to be told twice. He shifted the transmission into drive and the convoy left as fast as it could.

With the convoy leaving the rest of the company, the pilot had a choice to make. He could stay and decimate those who had remained behind or he could follow what appeared from the air to be five Iraqi tanks leaving their staging area and on their way to kill Marines.

Aber watched as the A-10 followed the medevac convoy. *Please realize who those guys are and leave them alone,* he silently begged the pilot of the A-10 as he watched it follow them into El Jasiph.

With the A-10 attacking his company, Aber hadn't noticed it before, but while they were being attacked from above the Iraqis had stopped firing at them, they had been just as surprised by the friendly fire incident as the Marines had been. It wasn't until after the A-10 had left and Iraqi bullets started racing through the air and Iraqi mortars were being lobbed in at them that Aber had noticed that they had even stopped.

He took cover next to his command track and looked towards the city anxiously.

"If Alpha and Bravo, or even just one of them, don't show up here soon, all they're going to find when they get here are a bunch of dead Marines. There's no way we can hold out against this, not as weakened as I am in manpower now," the company commander mumbled under his breath.

Realizing just how badly Charlie Company needed help from Alpha and Bravo Companies, or at least some air cover, Aber turned around and went back into his command track. He hoped, probably beyond hope, but he hoped nonetheless, that he would be able to get through to someone on the Battalion TAC.

CHAPTER THIRTY-THREE

Gunnery Sergeant John Ganton, of Alpha Company, hugged to the houses of El Jasiph as tightly as he could as he ran along the city street. He had been ordered by his Platoon Commander to secure another alley with Private First Class Jason Campbell, who ran five feet behind him. They made as small of a target of themselves as they could as they ran, but bullets still bounced off of the buildings all around them.

As he ran, Ganton saw an Iraqi soldier wearing civilian clothing step out of a courtyard up the road from him, fire his AK-47 wildly and disappear back behind the courtyard wall. He stopped running and took a kneeling firing position. When the Iraqi stepped out from behind the courtyard wall again to fire, Ganton fired his weapon, hitting the Iraqi in the side of his chest. The enemy soldier fell to the ground. He died without even knowing that the man who shot him had even been there, but that is the nature of urban warfare.

With the Iraqi dead, Ganton and Campbell kept running, hoping an Iraqi soldier wouldn't end their lives with the same surprise that Ganton had ended the life of the soldier he had just shot. Ganton knew there was a very real possibility he would die here in El Jasiph on this very day, but if he did, he at least wanted to see the man who killed him.

After a few minutes of hard running, the two Marines made it to the alley they'd been ordered to secure. Once there, they found a young Iraqi woman carrying a basket of something.

Campbell pulled his weapon up to his shoulder but Ganton barked, "Lower your weapon! She's not a fighter!"

"How do you know Gunny?" Campbell asked, it was an honest question and not a challenge. He still didn't lower his weapon.

"She hasn't tried to kill us!"

"But we don't know what she's got in that basket!" Campbell said.

"True, and we won't either, now, lower your weapon, Private!"

Hesitantly, as if it were against his better judgment, Campbell did as he was ordered and the woman picked up her pace, quickly disappearing down an adjoining alley.

The two Marines took up a kneeling position with their backs to each other so that no one could sneak up on them.

Without realizing he said it, Ganton said, "I really hate this Peek-A-Boo, I Shoot You bullshit."

"What do you mean by that Gunny?" Campbell asked.

Ganton kicked himself mentally. He really didn't want to engage in conversation with Campbell because he knew that the Private First Class was looking to him to know all of the answers. What the younger Marines in the company didn't seem to realize was that this was his first war too and he was just as scared as they were. He might be scared, but he wouldn't let himself show it, if he acted scared, then the younger Marines around him would really be scared and that would do nothing but get good men killed.

In addition to not wanting the younger Marines to see his fear, seeing Corporal Troy Jandris survive being shot in the head had shaken him more than he had ever imagined anything could. If something like that happened to him, he hoped that it killed him outright. The last thing he wanted was to have to live life with a brain injury or, even worse, as a vegetable.

"It's just that I never imagined war being fought in a city," he answered. "My grandpa was a World War Two Marine who spent the war island hopping. My dad was a Vietnam Marine. From listening to them, I always thought of war as being something that was only fought in jungles or by storming beaches with entrenched enemy soldiers shooting at you. When I got my orders for Iraq, I imagined we would be fighting Haji in open desert, not on the city streets of some place that I never knew existed until we got here. This here, this reminds me of a high stakes game of Peek-A-Boo, that's why I called it Peek-A-Boo, I Shoot You."

Campbell laughed and then fired his M-16 down the alley. Ganton turned quickly to face the direction that Campbell had just fired and saw an Iraqi soldier with his arms wrapped around an obviously painful stomach wound.

"Finish him off," Ganton ordered.

"Why, Gunny? He's nothing but a stinkin' Haji," Campbell asked. This time the question was a challenge.

"Would you like to be left to die like that?" Ganton asked.

"No. Why?" Campbell asked.

"He might be a Haji, but he's still a man and I'd bet anything that he doesn't want to be left to die like that either. Do the right thing and put him out of his misery," Ganton ordered.

A few seconds passed and Campbell pulled his trigger again. This time the bullets hit the Iraqi in the back of his head, which reduced his head to a bloody pulp, but he was still in death. His suffering was over.

The gunnery sergeant turned back the way he had been facing originally.

"Why did you laugh about shooting him?" Ganton asked.

"What? Oh, no, Gunny, I wasn't laughing about that. I laughed about what you said and then he came running out of one of the side alleys and pointed his AK at me," Campbell answered.

"What did I say that was so funny?" Ganton asked.

"The whole Peek-A-Boo thing, I hadn't thought about it like that, but you're right. This whole thing is like a Peek-A-Boo game from Hell," Campbell answered.

Ganton saw another Iraqi appear from an alley on the other side of the street with an RPG launcher. He pulled his trigger and the man fell dead without having had the chance to fire his RPG at any Marines.

Peek-A-Boo, I shoot you mother fucker, Ganton thought when the man didn't move again.

CHAPTER THIRTY-FOUR

The Marines from Bravo Company who had attempted to support Charlie Company were having troubles of their own. While none of them had been killed yet, and only a few minor wounds had been taken, they were receiving heavy enemy fire from a Mosque that was loaded with Iraqi soldiers firing at them with AK-47s, RPGs and mortars. Luckily for the Marines, these Iraqis were worse shots than the ones that were decimating Charlie Company.

Private First Class Ron Faber had abandoned the Humvee that he had manned the fifty-caliber of just seconds before it was hit by an Iraqi mortar and destroyed. Though he had only been feet away from the explosion, Farber had been lucky to have only suffered a few light burns and nothing more. Now that he laid on a sand berm firing his M-16 at the Mosque, Faber really didn't even notice the burns.

On Faber's left was Staff Sergeant Dave Ligget and to his right was Lance Corporal Glenn Utter. All three of their faces were caked with sand and sweat. They each aimed their weapons at the doors and windows of the Mosque, none of them could see the Iraqis inside, but muzzle flashes and RPGs seemed to come from every opening. The Marines fired and hoped to stop an enemy soldier from firing anything again.

Another American mortar was launched and flew through the air. Faber, Ligget and Utter watched as it landed on the roof of the Mosque.

"How many is that now?" Utter asked no one particular.

"Haven't been counting, but it's gotta be close to thirty that have hit the roof," Ligget answered.

The Mosque was a massive building and the American mortars had blown several holes in the roof and were causing havoc both on the roof and inside, but the problem was simply that it was too big of a building. Despite so many mortars falling on the roof, Iraqi soldiers were still up there, lobbing mortars of their own at the Marines.

"You got kids don't you, Staff Sergeant?" Faber asked as he loaded a new clip of ammunition into his M-16.

"Yeah, I do. Why?" Ligget replied.

"What are you going to tell them about fighting here?" Faber asked.

"What?" Ligget asked.

The incoming Iraqi fire picked up, so for the next several seconds none of the Marines said anything. Instead, they fired back at the Mosque.

An Iraqi mortar landed several feet behind them, too far away to have been any danger to the three Marines. All it did was rain down what seemed to be an avalanche of sand on top of them.

"What are you going to tell your kids and grandkids about fighting here?"

"Haven't really had time to think about it. So far the only thinking I've done about my family is how much I hope I don't fuck up and make my wife a widow and leave my kids without their daddy," Ligget replied.

Ligget fired his M-16 at the Mosque as he thought about Faber's question. He had to admit, the thought of surviving to tell his children, and someday, many years down the road, grandchildren about fighting in Iraq intrigued him. Before his grandfather passed away, Ligget had loved listening to his stories of fighting the Japanese during World War Two.

"I suppose I'd steal Bill Murray's line from Ghostbusters," Ligget said after a few minutes.

"What?" Faber asked.

"You asked what I'd tell my kids about fighting here. I think I'm going to steal that line Bill Murray said," Ligget answered.

"Which one?" Faber replied.

"We came, we saw, we kicked their ass," Ligget said.

Faber laughed. "I like that."

Captain Dan Earl, Bravo Company's Commanding Officer, observed the Mosque from an open hatch inside his command track.

"Any luck yet, First Sergeant?" he asked.

"No, sir, the Battalion TAC is still bogged down," Bravo Company's First Sergeant, Warren Backer, answered.

Earl had had Backer working on trying to get some air support to throw missiles into the Mosque ever since they started receiving fire from it. So far he had had the same luck trying to get through that Charlie Company had been having.

"Damn it! I want to see that place blown to pieces. You gotta love when things go to Hell," Earl said.

"Yes, sir, wouldn't be war if things went as planned," Backer said with his normal desert dry sense of humor.

CHAPTER THIRTY-FIVE

The A-10 followed the medevac convoy from Charlie Company, like a vulture keeping an eye on an Old West pioneer who had foolishly drunk all of his water long before arriving at his destination.

Unaware that death stalked them from above, Private Burt Farrow lay on his back in the lead track, among the other broken bodies of wounded Marines. Thanks to the Morphine that Petty Officer Third Class Craig Trevino had given him, he didn't feel any pain from his shoulder, which would require hours of surgery to bring back, even remotely, close to normal function.

Trevino moved from Marine to Marine, giving each of them the best medical care he could under the circumstances. Having already done everything he could reasonably be expected to do, Trevino mostly just kept a careful eye on each of them and hoped he could keep them alive long enough to make it to the doctors at the battalion aide station across the Euphrates River before he lost anyone else.

As Farrow lay on his back, lost in his own little, high as a kite, world, the A-10 soaring above them sighted in on the track that he was in. Everyone in the track was blissfully unaware of the immediate danger they were all in.

High above them, the A-10's pilot locked a Maverick missile onto the lead AAV and fired. The missile raced through the air and stuck the lead track on the left side of its rear cargo hatch. The explosion pushed the AAV forward, kind of like someone who is goosed by someone expectedly jumps forward, and destroyed the cargo hatch's locking mechanism. The hatch lowered and dragged behind it as the wounded track pushed on, creating a curtain of sparks as it did.

As frightful as the explosion had been, it could have been much worse. No one in the track had been wounded any worse than they already were, so the only real damage that the A-10's missile had caused was to the cargo hatch and the jangled nerves of those inside. No one had been killed.

The A-10 over flew the medevac convoy and, still believing that the vehicles that he was firing at were Iraqi tanks, the pilot brought his aircraft around for another pass at them. From inside, the Marines couldn't see the Air Force's tank killer coming in for a second attack run.

This time the Warthog pilot achieved missile lock on the fourth track. As soon as he had the AAV in missile lock, the pilot squeezed his trigger and let another Maverick fly.

Until his dying day, Corporal Marcus Andrews, the Marine who had jumped into the track commander's seat, would have no idea what caused it, but he had a sinking feeling in his gut that he needed to bail out of the vehicle. It wasn't a "something's wrong so I'd better pull over and find out what it is" feeling, it was a panicked "get out of here now or you're dead" feeling. Corporal Andrews was not the track's commander; there had been so much chaos at Charlie Company's position that very few of the tracks' original drivers and commanders had been included in the convoy.

Without really realizing he'd done it, Andrews opened the door and jumped out. The Marine in the driver's seat had the time to do little but turn his head towards where Andrews had been and begin to stare at the empty seat slack jawed before the A-10's missile struck.

Andrews hit the ground and rolled to his back just in time to see the Maverick strike the track he had been in just seconds before. The missile hit and the sides of the track bulged out in the split second before the explosion split it open. The explosion spewed forth Marines who were on fire like a volcano spews out fiery balls of magma when it erupts.

Oh my God, Andrews thought as he saw the explosion. Deep down inside of him, his gut told him that, out of everyone who had been in the track, he was the only survivor. The fact he had miraculously escaped severe injury when he bailed out of the track, while it was still in motion, hadn't struck him yet.

Everyone in the fifth track saw the fiery death of the AAV in front of them. The sight was met with all sorts of shocked exclamations, depending on each Marine's personality and how they respond naturally to the unexpected.

Some of them yelled for the driver to stop so they could rescue anyone who might have survived the explosion. The flaming wreck of the AAV was between them and Corporal Marcus Andrews, so no one even saw him.

"Look at that. They're all dead. No one could have survived that. Stopping will do nothing but paint a big target on us," the Marine in the driver's seat said to everyone who was yelling at him to stop.

Everyone realized he was right and an eerie silence fell over the Marines in the fifth track. They didn't know where the A-10 was, but they knew that it was up there somewhere, circling them like a great white shark circling swimmers in the ocean, just waiting to pick its next target.

Manny Alverez returned to where he had been sitting and silently prayed that God would spare the track he was in, the track that Irvin was in and the track that carried their wounded friend, Burt Farrow. He didn't know that the track carrying Farrow had already been hit, just without the catastrophic results of the missile strike that he had witnessed.

The world as Marcus Andrews knew it was spinning. His heart dropped as the fifth track keep going without stopping to pick him up. Had he really narrowly avoided dying in the explosion only to die trying to make it across El Jasiph on foot? Andrews didn't know what El Jasiph was like on a normal day, but today is was like South Central Los Angeles on crack and he really doubted that a lone American trying to make it across the city all by himself would live too long.

He looked around and saw his M-16 lying on the ground next to a house.

I wasn't thinking about what I was doing, but at least I grabbed that before I jumped. Thank God for small favors, he thought as he walked to his weapon and picked it up.

Andrews quickly checked his weapon to make sure it hadn't been damaged and that the barrel didn't have a buildup of sand inside of it, which could cause the M-16 to blow up in his face.

Well, things are starting to look a little brighter, Andrews thought when he was that his weapon was still in firing condition.

A moaning noise from the rear of the destroyed track caught his attention and Andrews's mind instantly jumped to George A. Romero's Night of the Living Dead.

There you go genius, Andrews thought, *let your imagination get the best of you here of all places. No zombies here in scenic downtown El Jasiph, just a bunch of Hajis. And Hajis die a lot easier than zombies so you got one of the long straws this time.*

Zombie or not, Andrews wasn't going to let whatever was making that moaning sound take him by surprise. He knelt, brought his weapon up to his shoulder and pointed it in the general direction that the sound was coming from.

Seconds, which seemed like years, passed and Andrews saw a man in a burned, but still recognizable, U.S. Marine uniform appear. The Marine's face was horrific to look at, everything from his top lip on up was so badly burnt that it really no longer looked human and the Marine's eyes were now nothing but charred, sightless orbs that were no longer good for anything except taking up space in the man's head.

That's gotta suck. Neither of us are going to survive, maybe I should just shoot him now and put him out of his misery, Andrews thought with the sight of his M-16 squarely on the blind Marine's chest, but, deep down, he knew that he could never shoot a fellow Marine.

Unsure of what to do, this wasn't exactly a situation that he had been trained for, Andrews continued to stare at the wounded Marine, who stumbled about blindly and moaned in his agony. It took a few seconds but he recognized the man as Private First Class Lee, "New Hickie", Newickki. They called him New Hickie as a play on his last name. It was a nickname that Newickki despised, but couldn't get the guys to quit calling him by.

"New Hickie!" Andrews yelled. "Is that you?"

Newickki's head started turning wildly in all directions as his panicked and pain wracked mind tried desperately to figure out where the person calling his name was.

"Who…who is it?" Newickki asked.

"It's me!" Andrews answered.

"Who the fuck is me?" Newickki yelled, annoyance clear in his pain-filled voice.

Andrews mentally kicked himself. *Great job dumbass, how is a blind man supposed to know who you are?*

"It's Andrews!"

"Andrews? Where are you man?" Newickki asked.

Andrews caught himself a split second before he said, "Over here." "Stay where you are, I'll come to you," he said instead.

With his weapon at the ready and hunched over to make as small of a target of himself as he could, Andrews ran as fast as his legs would carry him to Newickki. As soon as he got there, he put his hand on Newickki's upper arm and said, "I'm here."

"What the fuck happened?" Newickki asked.

"We got hit by a Warthog," Andrews answered.

"How many others?" Newickki asked.

"How many others what?" Andrews replied.

"How many others made it?"

Forgetting that Newickki couldn't see him, Andrews shook his head sadly. "It's just us."

"Don't know how lucky we are," Newickki replied grumpily. "You all fucked up like I am?"

"No, I'm in pretty good shape," Andrews answered.

"Good, then you get to lead the way," Newickki said.

Despite their situation, the comment made Andrews chuckle.

Newickki put his left hand on Andrews' shoulder and Andrews brought his M-16 back up to his shoulder.

"This really hurts brother," Andrews heard Newickki say from behind him.

"I bet. Don't worry, we'll find a corpsman and get some good drugs into you," Andrews said as the two men started making their way, very slowly, through the streets of El Jasiph.

Still believing he had a column of Iraqi armored vehicles below him, the A-10 pilot circled around again for another attack run on the medevac convoy. As he sighted in on the fifth track, the pilot became concerned about his ability to take out the vehicles below without unnecessarily endangering the civilians of El Jasiph, he'd done all he could to thin their number out and now it would be up to the boots on the ground to take care of what was left.

Instead of firing on them again, the pilot pulled back on his control yoke, which caused the tank killing aircraft to climb high into the air, turned the A-10 around and started flying back to where he had originally started raining death down on top of Charlie Company.

When he arrived, the pilot saw that Charlie Company was still heavily engaged with Iraqi forces in the city. The scene far below him didn't make any sense, why would Iraqi forces fire on Iraqi forces, but from everything he had been told, there were no friendly forces in the immediate area and it was open season on everything beneath him.

He unknowingly locked a Maverick missile on to Charlie Company's command track and rested his finger on the trigger. That was when the call came in informing the pilot that he had been attacking U.S. Marines, not Iraqi soldiers.

Cursing himself, the pilot took his finger off of the trigger and pulled back on his yoke. Climbing high in to the sky, he turned his A-10 around and went in search of Iraqi targets.

In his command track, Captain Brain Aber still tried to raise Alpha and Bravo Companies on the Battalion TAC. He didn't have any idea that he had just been seconds away from an explosive death.

CHAPTER THIRTY-SIX

The smell of unwashed human bodies, blood, urine, feces and death hovered over Marines who had been left behind to guard the vehicles that had been mired in the Iraqi sewage trap. The stench was so powerful, so overwhelming, that many of them honestly wondered if they would be able to wash it off fully or if there would be a taint of the odor that would always hover over them.

The noise of weapons being fired, of RPGs and mortars exploding and of the unarmed Cobra attack helicopter making its borderline suicidal attack run assaulted the Marines' ears and added to the overall confusion of their situation. All in all, it was not a pleasant situation.

It would have been easy for the Marines of Bravo Company to have fallen into chaos. Only the fact that those in leadership positions were able to keep calm and think rationally in the heat of the moment kept that from happening. It wasn't that individual discipline would have broken down, as a matter of fact, considering their situation, the Marines who had been left behind were conducting themselves admirably. What those who led them did was coordinate their men's field of fire so that their individual efforts weren't haphazard and wasted.

Twenty-five- year-old Sergeant John Davies ran through heavy Iraqi fire doing just that. The Marines Davies was responsible for were spread all over the place, some on dry land and others firing while standing or kneeling in the sewage that had caused their vehicles to become stuck.

Mankind is born with an innate sense of self-preservation, in that more often than not, we will do whatever it takes to preserve our lives, to see one more day. This was not the case for Davies, not that day. He ran from position to position, through dry sand and the thick mud caused by the sewage, without giving any heed to the Iraqi bullets that struck the mud brick and cinder block walls around him. He ran through the bullets that were zipping through the air with no more concern than he had had when he played in the rain as a child.

It might sound strange to those who had never been in his position, but at that moment in time, Davies didn't care if he was shot, he didn't care if he was blown up and he didn't care if he died. All that he cared about was making sure his men did their job and did it right, he was a leader and his job was to lead. If it meant this was the day he died, right here on the sandy, unpaved streets of El Jasiph or face down in the sewage sludge that mired their vehicles, then so be it. The Marines had trained him to lead men in combat and gave him the responsibility to do exactly that. He couldn't fathom himself doing anything less.

Aside from his own safety, one thing that Davies gave little attention to was the progress that the tank retriever was making in pulling the mired vehicles from the sewer mud. He hadn't really given it much thought at all until he happened to see the tank retriever pull one of the M1A1 Abrams completely free of the sludge.

As soon as the tank was free of the sewage, its main gun sighted in a building and fired. Its shell flew through the air and hit the building it had aimed at right in between floors. The M1-A1 fired again, but by the time the second shell arrived, the building was nothing more than a pile of rubble.

The sight caused Davies to take his attention off of making sure his men were firing at the targets that they should be firing at and he looked up at the sun.

We don't have much daylight left, he thought. *They'd better pull those loose before the sun goes down or we are well and truly screwed.*

Like the vast majority of the Marines on the ground in El Jasiph that day, Sergeant Davies had seen the movie Black Hawk Down and he was scared to death of finding himself stuck in the city after dark.

CHAPTER THIRTY-SEVEN

Sergeant Henry Quartermaine, one of the snipers assigned to Alpha Company, stood on the roof of a different building, bringing down one Iraqi soldier at a time. He found it disheartening when their tanks rolled away, but he had understood why they did. There simply weren't any targets they could hit from where they had been, so they left to hit the Iraqis from the flanks. Still, though, it gave a psychological boost to the Marines to have the M1-A1's here with them and gave the Iraqis a reason to think twice before revealing their position by firing at the Americans.

Iraqi after Iraqi fell dead from Quatermaine's shooting. It didn't matter if they were in a window, hiding behind a courtyard wall, or if they were firing from a rooftop, Quartermaine was a superb marksman and could drop someone no matter where they were.

The Iraqis seemed to grow increasingly bolder with each passing minute, so he didn't suffer from a lack of targets. Quartermaine himself wasn't under any direct fire, which made it easier for him to drop those who were firing at Alpha Company, but he found himself wishing more and more that the tanks would return. It might have just been the almost unbearable heat playing tricks on his mind, but it seemed to him that as the Iraqis' confidence increased, the Marines' confidence decreased. Regardless, if it was just how he perceived things or if that was how they really were, Quartermaine didn't like it.

Alpha Company had been ambushed and, while they were fighting back against it very well, Quartermaine knew that it was only a matter of time before things took a turn for the worse and when it did, it would be really bad.

Over the noise of the fighting around him, Quartermaine heard what sounded like thunder rapidly approaching Alpha Company's position. His heart jumped with excitement when he saw the four M1-A1 Abrams that had been assigned to Alpha Company returning.

The sight caused him to smile. It had the instant effect of making him feel better about their situation. He allowed the crosshairs of his scope to fall on an Iraqi soldier who was taking aim at the tanks with an RPG and pulled the trigger. His bullet hit the Iraqi right below his throat and the man fell. Quartermaine observed the spot where the Iraqi fell for a few seconds longer. He believed that he had fired a kill shot, but his training and taught him to make sure his target was dead before he started searching for his next one.

That was when he finally started receiving enemy fire and it came heavy. He didn't know where it was coming from, but an Iraqi with a heavy machine gun had either seen his muzzle flash or had actually seen him and taken aim. Whichever it was didn't matter; if he tried to continue sniping Iraqi soldiers from his current position, he would join the day's dead and that was something he really didn't want to happen.

Quartermaine dove to the roof to avoid being hit. With his M40-A3 sniper rifle in hand, he crawled to the stairs. Quartermaine descended the stairs and moved to the window, in what looked like a bedroom, and watched for the muzzles flashes of the heavy machine gun that had nearly ended his life.

The Iraqi apparently hadn't realized that Quartermaine had moved because he continued to pour a tsunami of bullets at the roof of the building the sniper was in. That made finding where the enemy soldier was a really simple matter. The Iraqi was well concealed by the shadows inside a building right across the street.

Since he couldn't see the Iraqi soldier, Quartermaine placed his scope's crosshairs just above the muzzle flashes and pulled the trigger. Seconds later the heavy machine gun fell silent.

With his life no longer in immediate peril, Quartermaine looked back towards the tanks and saw that the tank commander was standing out in the open talking with Lieutenant Jim Everett, the Battalion Commander.

Quartermaine couldn't tell what the two officers were saying to each other, but they were both pointing at various buildings so he assumed that Colonel Everett was pointing out the ones that were proving to be particularly troublesome.

Curiosity got the better of him and Quartermaine kept looking over at the officers after every couple of shots that he took. After a few minutes, he saw the tank commander get back into his tank and a few seconds later, two of them drove off.

The tanks reappeared on the other side of Alpha Company's position after a few minutes and, in what was obviously a coordinated attack, all four tanks began firing on certain buildings, quickly reducing them to nothing more than piles of rubble.

Once the tanks started firing, the Iraqis seemed to learn their lesson and the incoming enemy fire fell silent. When the tanks didn't have any more clear targets they could hit from their current positions, they drove off to once again take the fight to the Iraqi soldiers in other areas of El Jasiph.

As soon as they were gone, the Iraqi fire picked back up and Alpha Company's fight resumed.

CHAPTER THIRTY-EIGHT

Track five in the Charlie Company medevac convoy sped through the streets of El Jasiph as fast as the driver could force it to go without smashing into the back of track three, track four already having been destroyed.

Being the last track in the convoy, those in track five were the only ones to have actually witnessed the fiery death of track four and the sight drove home the fact that they weren't out of the woods. They wouldn't have any measure of safety until after they crossed the Euphrates and were on their way to the battalion aid station.

The constant sound of Iraqi bullets bouncing off of the AAV's hull was a constant reminder of that. The Marines on track five wanted to breathe a sigh of relief when it appeared that the A-10 had given up on wiping them off of the face of the earth, but they really couldn't because every now and then an enemy bullet punctured through the track's hull and came awfully close to hitting one of its occupants. So far no one had actually been struck by one of them, but there had been too many close calls and none of the Marines were under the delusion that bullets could keep punching through forever without some of them eventually being hit, possibly even fatally.

It wasn't their own safety that weighed most heavily on their minds, each of them had already accepted the fact that they would probably die in El Jasiph and that acceptance made it easier to do whatever needed to be done. Their chief concern was for each other and the already wounded Marines in tracks one and three. If they didn't make it home, they wouldn't know it, but the impact of each missing Marine, each friend and brothers in arms would be felt as new faces replaced those lost. Even worse than seeing new faces would be seeing the wives, girlfriends and children of those who had made the ultimate sacrifice. The Marines may have lost a friend, but the wives and children had lost their entire world, at least for a while until things began to return to a semblance of normal.   Manny Alverez was concerned mostly about his two best friends, Burt Farrow, who was wounded and on track one and Andy Irvin, who was on board track two. He had wanted to be on the same AAV as

Irvin, but there were so many men in Charlie Company jumping on board that there hadn't been any room for him, so he'd found a place on track five.

None of the Marines who had gone with the medevac convoy had done so out of cowardice. It wasn't the promised safety waiting on the other side of the Euphrates River that had caused them to scramble for a spot on the convoy. The injured Marines were largely helpless to defend themselves and they'd gone along to make sure that the wounded made it to the battalion aid station. Once the wounded were safely in the hands of the medical staff, each of the uninjured Marines had every intention of returning to the fight, the idea of not returning to what remained of Charlie Company was just as unfathomable as letting the wounded make their way through the El Jasiph gauntlet by themselves.

Of all of the things that the Marines on track five were expecting to happen, none of them did. Lady Luck wasn't with them and the one thing that they knew could happen, but really didn't expect to happen, was what occurred. A rattling noise came from the engine, the AAV shook violently and the large, lumbering machine coasted to a stop.

"What the fuck man? Why are you slowing down?" one of the twelve Marines in the track asked.

"It's not me! The damn thing just died!" the driver protested.

"What's wrong with it?" Alverez asked.

"How am I supposed to know? I'm a grunt, not a freakin' grease monkey. I'm lucky if I know a lug nut from your sister's pussy," the driver replied, obviously annoyed.

AAVs were notoriously unreliable. They had been designed for amphibious landings and not for urban desert warfare. Even then, they had an extremely bad habit of dying in the water without even making it to the beach to deploy the Marines they carried.

The driver's announcement was met with various moans, groans and curses of dissatisfaction. None of which were aimed towards the driver, no one blamed him, they were counting on a piece of junk to get them through El Jasiph and they knew it. This time, the piece of junk just wasn't up to the job.

Staff Sergeant Nick Pacini, who was sitting in the AAV commander's seat, immediately jumped on the radio and started talking on the Company TAC, "Track five to all tracks in medevac. Track five to all tracks in the medevac convoy. We've broken down. Do not, repeat, do not stop for us. Get the wounded to the people that can help them and send someone for us when you can. We'll find somewhere to hole up."

"Are you sure?" a voice Pacini recognized asked.

"Damn it, York! Yes, I'm sure! If you ask again I'm going to kick your ass!" Pacini replied.

A fist size lump, which felt like a rock, sank into the bottom most pit of Pacini's stomach as track five slowed and the rest of the convoy left them in their dust. *I hope that decision doesn't end up putting us in a Custer's Last Stand type of situation,* he thought as the amount of enemy bullets hitting the track seemed to pick up. As soon as track five rolled to a complete stop, Staff Sergeant Pacini stood up and looked at the guys in the back. Their situation was bad and Pacini knew it, but he managed to look a lot calmer than he felt. "Ok, guys, we're officially broken down. What I'm going to do is zero out all radio equipment. The rest of you, grab all the ammo you can carry. To our right is a house that's door is lined up almost exactly with our cargo hatch. As soon as the hatch lowers, run harder than you've ever run and get inside that house. Be ready to fight, we don't know if there are any Hajis in there or not."

Zeroing out the radios meant to delete all American frequencies from them so that, if they were to fall into Iraqi hands, the enemy soldiers couldn't use them to listen into the Marines' conversations.

He looked at each of the Marines. "Any questions on that?"

There weren't. "Ok, get to it then. I want to be out of here in under a minute."

Each of them set about doing what they needed to do in order to evacuate track five. As soon as everything was ready, the driver hit the button that lowered the cargo hatch and the Marines ran out, just as they would have if they had been storming a beach in some other part of the world.

Their flight from the track had been expected and as soon as the Marines started pouring out, the amount of incoming Iraqi fire turned frenzied. Four of the Marines were hit by enemy fire as they abandoned track five and were either carried or dragged to the house they were retreating to by the others.

As soon as the Marines made it inside, they encountered a woman in her mid to late twenties and two young children. Pacini saw a few of the Marines point their M-16s at them.

"Shit! Civilian! That's the last thing we need to be responsible for!" the staff sergeant thought but said, "Lower your weapons! Lower them, now goddamnit! They're civilians!"

Pacini looked around. "Secure the rest of the house, I'll stay here with them. Alverez, find a room on this floor that would make a good casualty collection point until we get picked up."

He turned his attention to the Iraqi civilians. "And what am I going to do with you guys? No matter what I decide, it might get you killed and that really bites," Pacini was speaking to himself because he didn't imagine that anyone in the family spoke English.
Bullets hit the house at such a rate that it sounded like it was raining heavily outside. The heavy incoming fire died out after just a few seconds and Pacini decided what to do with the family.

Motioning towards the door with his M-16, the staff sergeant began yelling, "Out! Go! Get out of here! Now!"

Even though she didn't speak a word of English, the mother apparently understood what he was saying. She wrapped her arms around her children protectively and all three of them ran for the door.

Once they were gone, Pacini was surprised at how relieved he felt to have them out of the house. Things were likely to get bad and the last thing he wanted on his conscious, if he survived, was the death of an innocent woman and her children.

Less than a minute later the others returned from clearing the house.

"We're clear, Staff Sergeant," Corporal Stan Fackler reported.

The incoming fire turned heavy again.

"Alverez, did you find a room that would make a suitable casualty collection point?" Pacini asked.

"Yes, Staff Sergeant," Alverez answered.

"Good, I want you and Newtson to move our wounded there. There is only one way in and out of this house so I want you two to defend that door with everything you have."

Neither Marine replied verbally, but they both nodded their heads in acceptance of their orders.

"The rest of you, with me," Pacini said, already heading for the stairs.

Leaving Private Manny Alverez and Lance Corporal Tim Newtson behind, everyone else followed Staff Sergeant Nick Pacini up the stair

Once they made it to the roof, Pacini went to work arranging the six Marines with him to form a three hundred and sixty degree defensive position. With four Marines wounded and Alverez and Newtson still downstairs, his available manpower was cut in half. He didn't like the situation, but he would have to make do with it.

Downstairs, Alverez and Newtson finished helping their wounded comrades into the room that was least likely to have bullets find their way into it. They had the wounded laying close to each other.

"I don't like this," Newtson said as an Iraqi bullet ricocheted into the room. "They might not get shot directly, but some of these guys are still going to get hit."

"Doc! Get me Doc Trevino!" one of the wounded Marines called out. He had been shot in his left side and Alverez, who was exhausting his small and very basic first aid kit bandaging up the wounded while Newtson tried to figure out how to strengthen their casualty collection point hadn't been able to find an exit wound.

"He's not here. He is on track one with the guys who needed him the most when we loaded up," Alverez replied, trying to have a good bedside manner despite everything that was happening.

Still trying to create a greater degree of safety for the wounded, Newtson ran from the room.

"Shit this hurts," the Marine who had been shot in the side said. "Give me something man."

Alverez shrugged helplessly. "I wish I could, I really do, I don't have anything. You're just going to have to tough it out until we can get picked up."

"Fuck you, you cock suckin' motherfucker! This shit hurts!" the wounded Marine replied.

Wanting to punch the man, Alverez forced himself to keep his cool. He probably would have been the same way if their positions had been reversed.

Before Alverez could say anything else, he heard Newtson yell, "Alverez! Come here!"

Alverez ran to see what Newtson needed and found him looking out one of the front windows.

"What?" Alverez asked.

"Look at that."

"Look at what? There's a lot that's out there."

Newtson pointed at sandbags, which had been an Iraqi machine gun nest before being abandoned by the man who had manned it. "The bags, we could get those and stack them around the shot up guys."

Unable to believe that Newtson was actually suggesting that they go back out there, Alverez looked at him. "Are you nuts? We wouldn't make it five steps before some Haji gunned us down."

"I'm going for them. Cover me," Newtson said, already running for the door.

Alverez looked on incredulously as Newtson ran through heavy enemy fire to the sandbags and then back carrying three of them. He was too stunned to even fire his weapon.

When he made it back to the house, Newtson didn't even slow down until he reached the wounded men. He placed the sandbags around them and ran back out.

"Are you out of your mind?" Alverez yelled as Newtson ran by him and again went outside. Not as stunned this time, he had the presence of mind to point his M-16 out of the window and give Newtson cover fire as he made his second trip to the abandoned machine gun nest and returned with three more bags.

The Marines on the roof were just as stunned as Alverez when they saw Newtson running around outside and provided cover fire for him as well.

His adrenaline rush was the only thing that kept Newtson from feeling the weight of the sandbags as he ran. After making a third trip, Newtson decided that he had done all he could do for the wounded Marines by bringing nine sandbags in to help block bullets from hitting them.

With the casualty collection point as secure as it could reasonable be, Newtson returned to the front room and knelt to the side of the door and joined Alverez in firing at Iraqi soldiers.

CHAPTER THIRTY-NINE

Private Charles Schuman and Lance Corporal Dalton Crocket lay next to each other on a sand berm firing their M-16s back towards El Jasiph. Both of them had tried unsuccessfully to get a spot on the medevac convoy. At first they had been irritated over not having been able to get one, but after watching the A-10 follow the five tracks into the city, both of them were happy that it had filled up before they found a place. For all they knew, the Warthog had laid waste to the entire convoy before it returned to their position and started what looked like was going to be another attack run, but then flew away without firing on them again.

After the way his day had gone, Schuman didn't have any doubt that his God was there in the Iraqi desert with him. First, he had survived being in the track that was hit going through El Jasiph when so many others, including Sergeant Mark Valentine, who had been sitting right there with him, hadn't. Not only had he survived it, but he had escaped with nothing more serious than a few minor burns and some singed hair. As if that hadn't been enough, then he had not been able to find a spot on the medevac convoy, and his gut told him that not all of those guys were going to make it across the Euphrates. If the A-10 pilot hadn't taken them all out, the Iraqi soldiers in El Jasiph were bound to kill at least a few of them as they worked their way back through the city. Yes, his God was with him and that gave him confidence that his Lord was going to see him through this bloody day.

"You think the convoy made it?' Crocket asked.

"The A-10 wasn't gone too long. He may have taken some of them out, but I don't think he had time to get all of them," Schuman answered.

Crocket rolled to his back and pulled a pack of Basic brand cigarettes out of one of his pockets. He freed one from the pack, stuck it in his mouth and lit it. Instead of rolling back to his stomach and rejoining the fight, Crocket just stayed on his back, savoring the taste of the cigarette as he puffed on it.

"What're you doing, man?" Schuman asked. "If you're going to smoke that thing, at least shoot while you do."

"Fuck that shit, man," Crocket replied. "I haven't had one of these in hours so I'm taking me a little smoke break. Those camel fuckers over there can wait a couple minutes to die."

Schuman's clip ran out of ammunition so he rolled to his back, ejected the empty clip from his M-16 and loaded a fresh one. Crocket held his pack of cigarettes out to him and asked, "Smoke?"

"You know I don't smoke," Schuman replied as he rolled back to his stomach and resumed firing his weapon towards El Jasiph.

"Maybe you should start. You'll live longer," Crocket said.

This time Schuman didn't even bother to reply. He knew Crocket was just trying to get a rise out of him.

"You not smoking, that's a religious thing, right?" Crocket asked as he took the last drag off of his smoke, flicked it away from him and rolled back to his stomach.

"Sure is," Schuman answered without looking at him. "Doesn't the Bible say some pretty bad things about killing people? Why did a nice Christian boy like you join the Corps anyway?" Crocket asked, the question was genuine and not meant to be mocking.

To Schuman, it seemed like an odd time for someone to be asking him questions about his faith, but then he could fight and answer at the same time. Still firing his weapon and not looking at Crocket, Schuman answered, "The Bible tells us how we should conduct ourselves in our personal lives. It doesn't refer to military service, so when it says 'thou shalt not kill,' it's telling me I shouldn't kill my neighbor because they have something I want or someone who happens to make me mad, but it does not mean that Christians should not serve their country and fight in a war if that's what their military service ends up meaning. God's people, throughout history, have always taken up arms when it was necessary."

"Gotcha. I always meant to ask you about that," Crocket replied, firing his M-16.

Several minutes went by without either man saying anything. Crocket broke the silence by saying, "I'll tell you one thing," but then he didn't say anything else.

"What's that?" Schuman asked.

"If Alpha and Bravo don't get here soon, I think we are all going to be seeing your God face to face here shortly."

Schuman stopped firing and looked at Crocket. "That's certainly a possibility. I'd much rather die a really old man with my wife, children, grandchildren and great grandchildren surrounding me, but if I die here today, at least I know where I'll wake up."

Both men went back to firing at the Iraqi soldiers in the city of El Jasiph.

A couple of minutes later, Crocket stopped firing and looked at Schuman. "You said you know where you'd wake up if you die here, right?"

"Yup," Schuman answered without looking at him. It was a question he could answer while still firing his weapon.

"How about me? If I die here, do you know where I'll wake up?"

Schuman made it a point not to try to make people who didn't want to hear about his religious beliefs listen to him. At the same time, he was always happy to share his faith with those who asked him about it. He wasn't ashamed of what he believed. However, this wasn't the time or the place. He needed to wear his Marine hat and not his missionary hat in the firefight they were engaged in, just as he felt that Crocket needed to wear his Marine hat and not his potential convert hat. If they didn't, then one or both of them could easily make a mistake that would result in them waking up in eternity.

He stopped firing his weapon and looked at Crocket, who was still looking at him, his eyes begging to be told that he was in a good place spiritually. He seemed to be convinced that he wasn't going to make it through the day. Schuman wouldn't lie to the man, but this simply wasn't the time to get into a discussion on the requirements that God has placed on humanity if they wish to spend eternity with him in Heaven.

"Tell you what," Schuman said. "You just focus on not getting shot or blown up, focus on killing Iraqis and not letting them kill you. When this whole mess is done and we are someplace where we can talk about this safely, we will. Deal?"

"Deal," Crocket replied.

Both men rejoined the fight and the topic of religion didn't come back up for the rest of the battle.

CHAPTER FORTY

Corporal Marcus Andrews and Private First Class Lee, "New Hickie", Newickki, the only two of the twelve Marines on track four to survive its destruction, made their way slowly through the streets of El Jasiph on foot. Andrews had his M-16 up to his shoulder as he walked, moving his upper body as if it were all one piece that didn't have a choice but to move as a solid unit. His eyes didn't look anywhere that the barrel of his weapon wasn't facing. Newickki, blinded and with his face covered by third degree burns, held onto Andrews's right shoulder with his left hand, trusting him not to get them both killed.

Throughout his life, Andrews had experienced a lot of stress, but none of it had been anything like this. His blood pressure was so high it felt like his heart could burst through his chest at any moment. His profuse sweating was not just due to the heat, although admittedly that did play a big part of it, and he hadn't even breathed as hard as he was walking through the streets of El Jasiph when he had been going through recruit training at Parris Island.

An Iraqi woman walked out of a house in front of him and, in the split second that he had to make the decision to pull the trigger or to hold his fire, he realized that she was an unarmed civilian and not a threat to them. Instead of killing her, Andrews let go of his weapon with one hand and motioned her back inside. Even with the language barrier between them, she understood what he was saying to her and she disappeared back into the house.

Once she was out of sight, Andrews and Newickki moved forward again, at a pace that felt painfully slow. Five minutes later, an Iraqi soldier in civilian clothing came running out of a house three houses in front of the two Marines.

Andrews let his weapon's sights fall on the man's upper torso and pulled the trigger. The Iraqi fell to the ground with a sucking chest wound and instinctively Andrews wanted to put the enemy soldier out of his misery. They might be on opposite sides of this war, they might be trying to kill each other, but he was still a man and a sucking chest wound was a bad way to die. Andrews knew that if their roles had been reversed, he certainly would have hoped that the Iraqi soldier would put a bullet in his head.

He had just started to move towards the dying man when another Iraqi soldier came running out of the same house. The first one hadn't known that Andrews was there, but the second one did and when he exited the house he was turned facing the two Marines and came out firing.

The feel of bullets brushing by his face, yet not hitting him, sent a jolt of adrenaline through Andrews that was stronger than anything he had felt before and he pulled his trigger again. His bullets hit the Iraqi in the chest, throwing the man backwards, as if he had been attached to invisible ropes that had pulled him.

"Down, New Hickie!" Andrews ordered.

Newickki didn't say anything and, even though he could still feel the injured man's hand on his shoulder, the lack of a reply concerned Andrews because of how severe Newickki's wounds were. A quick glance over his shoulder convinced Andrews that he didn't have anything to worry about and both men knelt.

"What's going on?" Newickki asked.

"Two Hajis just came out of a building ahead of us. I don't want to walk in front of it until I'm reasonably sure there aren't any more in there." Andrews answered as he took in the entire building, the door, the windows and the roof.

Several tension-filled and agonizingly slow minutes passed, but no one else came out of the house and Andrews couldn't see anyone else moving around inside or on top of it.

"Ok, New Hickie, let's get going," Andrews said.

"Lead the way," Newickki replied, not intending to refer to the fact that he was incapable of seeing what was ahead of them.

The two Marines started moving forward again. When they passed by the Iraqi soldier with the sucking chest wound, Andrews stopped to put the man out of his misery, but he had already died.

A few minutes later another woman walked out in front of them and Andrews made the split second decision that she wasn't a threat to them either and motioned her back inside her house. She complied just as quickly as the first one had.

Andrews and Newickki passed the woman's house and the next thing he knew, something pushed Andrews forward and he heard the sound of a weapon being fired from behind him. As he fell, Andrews turned so that he was facing the way he had just come from. In doing so he became separated from Newickki, and fired his M-16 wildly. His bullets struck an Iraqi soldier, who had come out of the same house that the woman had just gone back into, in the stomach, and he fell.

With the amount of adrenaline pumping through his body, Andrews didn't have any idea that one of the Iraqi bullets had hit him in the back and had become lodged near his spine. He mistook the blood he felt running down his back for more sweat, which already coated every inch of his body.

"Ok, New Hickie, let's find someplace safe. Things are getting just a little too hot out here for my taste," Andrews said. He wasn't referring to the temperature.

CHAPTER FORTY-ONE

On top of one of the houses, a young Iraqi soldier loaded another RPG into his launcher and sighted in on a courtyard wall that a group of three Marines from Alpha Company had taken cover behind. The young soldier didn't know if he had actually killed any Americans or not, but he really didn't care. His army had more people in the fight than the Americans did, so he didn't see any way the Iraqi Army could possibly lose this battle. He was enjoying himself. Whether he had actually killed anyone or not simply didn't matter to him.

He pulled his weapon's trigger and the RPG flew away from him, hit the courtyard wall and exploded. The explosion opened up a huge hole in the wall, but before he could load another RPG into his launcher, the young Iraqi had to dive to the roof because his RPG was being answered by hundreds of American bullets. None of the bullets hit him, but his right cheek was cut by a sharp piece of mud brick that had been broken loose and sent flying by the onslaught of incoming fire.

As soon as the Americans stopped firing at his position, the soldier popped back up to see if he could find out who had been shooting at him. If he could, he'd make sure that his next RPG killed them. He spent a few minutes observing the battle three stories below him, but the scene was so chaotic that he didn't have a prayer of figuring out who it was that shot at him. No one but Allah would have been able to tell one American from the next, so he vowed the next RPG he fired would kill at least one of them. He liked firing RPGs at the Americans, but up to that point, he hadn't come under fire and the young Iraqi soldier didn't like being shot at. With that personal vow made, he went to work loading his next RPG.

Lance Corporal Victor Lauralwood ducked into the CAAT Humvee as bullets fired from an Iraqi AK-47 bounced off of it. Once the storm passed, Lauralwood didn't waste any time getting back up to his Mark 19 grenade launcher. He had seen the window where the fire directed at him had come from and he responded by firing a stream of grenades through the window. Each grenade went off and no one else fired at him, or anyone else, from that window.

Corporal Bruce Hawkins hadn't moved from the Battalion Command Humvee since he took up position behind its fifty caliber machine gun early that morning. Lance Corporal Jack Tays fired his M-16 over the Humvee's hood, allowing the vehicle to stop the Iraqi bullets that were trying to end his life.

Neither of them knew where their Battalion Commander, the Battalion Gunner or the Battalion Sergeant Major was. Ordinarily that would have been a problem for them under their current circumstances, but neither of them really cared. The only concern either of them had at that point in time was still being alive once their first firefight was over with. Hawkins fired his weapon at an Iraqi soldier that had popped out from behind a courtyard wall and fired his AK-47 wildly instead of at any particular target. His large shells hit the Iraqi in his chest, which just disappeared in a gory, pink cloud. The man's lower body, arms, head and weapon all hit the ground, no longer connected to each other.

The noise of an approaching vehicle caught Tays's attention as he fired his M-16. Realizing that a firefight was a very bad place to take anything for granted, he quit firing his weapon and turned his head to see what was coming towards them. A few seconds later the first track in Charlie Company's medevac convoy came into view, its cargo hatch still dragged on the ground behind it, spraying everything it passed with a shower of sparks.

"Hey, Bruce!" Tays called up to Hawkins.

Hawkins was unable to hear him over the noise the fifty made when it was fired.

"Bruce!" Tays yelled as loud as he could manage.

"What!" Hawkins replied.

Tays pointed at the approaching AAV. "Check that track out man."

Hawkins turned his head to look in the direction Tays was pointing. "Wow! Something got them a good one didn't it."

"What do you think did it to 'em?" Tays asked.

"No clue, I'd guess they must've got hit with a glancing shot from an RPG or a mortar," Hawkins answered.

The US Marines weren't the only ones to hear track one's approach, the young Iraqi with the RPG launcher heard it too. The sound made him curious enough that he lowered the launcher from his shoulder and looked to see what was making it.

"Allah be praised," he thought when he saw track one. *"I'd really be able to brag if I kill one of those beasts."*

He brought the RPG launcher back up to his shoulder and pointed it at the track. The young man was smart enough to realize he would probably only get one shot at taking out the track and didn't want to squander it. He rested his finger on the launcher's trigger, but didn't pull it. When he took his shot, it had to count. He could kill a lot of Americans with one shot if he was patient enough.

Petty Officer Third Class Craig, "Doc", Trevino pulled the syringe that he had used to give Private Burt Farrow another dose of morphine out of the eighteen year old Marine's arm. Once he had recapped the needle, he allowed the syringe to fall to the floor of the track and looked around him.

Sooooooo many, the corpsman thought as he looked at the torn and broken bodies in track one with him. *War isn't supposed to be like this. It's supposed to be taking it to the bad guys and then coming home to ticker-tape parades and kissing all the honeys you can while looking dashing in your uniform with your chest full of medals for heroism. It's supposed to be the bad guys who are all shot up, not your guys.* His mind couldn't keep track of all of the Marines he had lost. He wasn't supposed to lose any, they get shot and he was supposed to come running in, through a wall of enemy fire, and save the day. None of them were supposed to actually die on him.

In just the few hours that the firefight in El Jasiph had been raging, reality had smacked Trevino in the face hard. War movies were one thing; real war was a completely different animal all together. Despite his best efforts, three more Marines had succumbed to their injuries and died on him since they had left Charlie Company's position. He was afraid to know how many had died on the other track that carried wounded. Charlie Company only had one corpsman and he was here, which meant the other track didn't have anyone to attend to the wounded.

Hundreds of tiny details regarding his patients ran though Trevino's mind. One detail that had escaped his attention was the fact that Private Farrow was sitting next to the AAVs ammunition stores.

Still on the house's roof, the young Iraqi soldier waited patiently for just the perfect shot at the track to open up. Deep down he knew Allah had delivered this group of Americans to him for him to kill and he would not disappoint his God.

He cursed as the thick black smoke trail of another RPG raced towards the AAV. If someone else robbed him of the kill that Allah had provided for him, the young Iraqi soldier vowed that he would find whoever it was and kill him in the slowest, most painful fashion possible.

The RPG hit the side of track one and exploded. The young Iraqi breathed a sigh of relief when it failed to penetrate the track's hull. The explosion caused the wounded vehicle to rock side to side a bit and probably shook the nerves of those inside even more than they already were, but the kill would still be his.

With his RPG launcher up to his shoulder, he waited patiently until the track lumbered to where it was right underneath him. That was when he noticed someone had failed to close the track's top hatch. He sighted in on the opening, pulled his weapon's trigger and the RPG raced away from him, trailing thick black smoke as it did.

Trevino checked the pulse of the Marine next to Farrow. It was growing weaker, but they were close enough to the Euphrates River now that he might just live long enough to make it to the doctors at the battalion aid station. At least Trevino hoped he would because he already had too many faces of the men he had failed to save that would follow him until his dying day. He really didn't need another one.

Things happened so fast that Trevino's mind didn't have time to register it when the RPG came through the open top hatch and landed on track one's store of ammunition. It exploded as soon as it hit and Petty Officer Third Class Craig, "Doc", Trevino and Private

Burt Farrow were killed instantly. They both went to their deaths without feeling any pain.

The explosion cooked off the rest of the ammunition that the RPG landed on and the initial explosion was followed by several more that tore the track apart.

Lance Corporal Tays's eyes grew wide in shock as he watched track one blow up. Without any conscious thought and without any concern for his own personal safety, he ran to the ruins of track one.

Corporal Hawkins covered Tays by firing at any Iraqi position that opened fire on Tays with his fifty-caliber.

The young Iraqi's heart was filled with joy when he saw the destruction he had brought to track one. His youthful enthusiasm overruled his commonsense and he allowed his RPG launcher to fall to the ground and started dancing. It was the only way he could think of to express his joy.

Hawkins saw the young Iraqi soldier dancing on the rooftop. Up until that point in time, he had fought dispassionately, he was there to fight and fighting was what he was going to do, it wasn't anything personal. That changed, however, when he saw the young Iraqi dancing with joy at the death of the AAV. The sight of an Iraqi dancing over the death of his fellow Marines filled Hawkins's heart with hatred and anger.

"I'll give you a reason to dance you camel cocksuckin' motherfucker," he mumbled as he placed his weapon's sights on the Iraqi soldier.

Hawkins pulled the trigger and sent a deadly stream of large, fifty-caliber bullets streaming towards the dancing Iraqi.

The young Iraqi was so caught up in his own excitement and joy that he had momentarily forgotten where he was and he kept dancing. When he got back to the small village he was from, this story was going to make him a hometown hero.

He didn't even see the fifty-caliber bullets coming at him. The first one hit him in the head, which disappeared from his shoulders and the rest of his body fell to the roof, landing on top of his RGP launcher. He died without even knowing it.

Lieutenant Colonel Jim Everett witnessed the destruction of track one also. "My God in Heaven," he muttered unconsciously at the sight.

He looked around him and saw Sergeant Major Art Jamison. "Sergeant Major, go find the corpsman. Tell him to start loading up Alpha's dead and wounded on a track." "Yes, sir," Jamison replied and then disappeared into the chaos that surrounded them. Everett's attention shifted to Chief Warrant Officer Pete, "Gunner", Lockhart, the Battalion Gunner, and said, "Get your ass over there. Get the dead and wounded off of that track and onto the track that Jamison is going to have Alpha's dead and wounded loaded on to. We need to get these guys to battalion aid now. I hope I don't need to say it, but the wounded get priority over the dead."

"Yes, sir," Lockhart replied and then started running towards the remains of track one, taking every Marine he came across with him.

Since they were still engaged in a heavy and sustained firefight, not every Marine from Alpha Company could be in on rescuing the dead and wounded from the destroyed Charlie Company track. Those who were part of the rescue operation needed someone to provide covering fire for them or they wouldn't stand a chance.

It took those involved in the rescue less than twenty minutes to load the dead and wounded from both Alpha and Charlie Companies onto one of Alpha's tracks. As it was about ready to roll out, Alpha Company's First Sergeant, Hector Urness, walked out of track one carrying a glob of melted human flesh that had solidified into a shapeless mass. "What's that," Sergeant Tom Daley asked.

"Part of a Marine," Urness answered.

"What part and who was it part of?" Daley asked.

"Hell if I know. Don't matter anyway. We don't leave any Marine, or any part of any Marine, behind. We take everything," Urness replied as he walked to the new medevac track and put the shapeless, nameless glob of human flesh onboard.

CHAPTER FORTY-TWO

"What have we gotten ourselves into, Staff Sergeant?" Corporal Stan Fackler asked Staff Sergeant Nick Pacini.

Pacini didn't respond right away. Instead of answering, he fired the M-203 grenade launcher, which was attached to his M-16. After his grenade hit the roof of the house he had aimed at, Pacini looked at Fackler and asked, "What're you talking about?"

"It just feels like we've found ourselves our own personal Alamo. Please tell me that we haven't," Fackler answered.

"We very well might have, but stop your bitchin' and get back to fightin'," Pacini said, pulling his M-16 back up to his shoulder.

"That might be easier said than done before too long," Fackler replied snippily.

"What do you mean by that?" Pacini asked, feeling his anger at the mouthy corporal beginning to rise.

"We're already running low on ammo. If we run out, we ain't gonna be doing much fighting," Fackler replied, refusing to be cowed by the extra stripes Pacini wore.

That was when Pacini realized for the first time that they were in fact running low on ammunition. "Didn't I tell you guys to grab as much as you could before you abandoned the track!" he barked.

"We did, Staff Sergeant. Problem is, we're shooting it off too quickly and we're just about out again," Fackler said.

Pacini looked at the other four Marines on the rooftop. "Everyone running low?"

"Yes, Staff Sergeant," one of them answered. "We have been for some time now."

"Keep shooting, don't let Haji know we're running low on ammo," Pacini said and started to head towards the stairs.

"What are you doing?" Fackler asked.

"I'm going to go get us more ammo," Pacini answered.

"What do we do if we run out?" one of the other Marines asked.

"Break off chunks of cinderblock and throw it at them. If they get too close, piss on their heads. I don't know, just keep their attention away from me," Pacini answered and then disappeared down the stairs.

The staff sergeant ran down the stairs as fast as he could without letting his forward momentum get carried away and send him tumbling down them. As soon as he reached the ground level, he took off in a mad dash to the front room and saw Newtson kneeling by the door and Alverez kneeling by a window. Both men were firing their weapons. Pacini came to a stop next to Newtson.

"Alverez, over here," Pacini said, his voice made it clear it wasn't a request. Alverez hurried to comply.

"Ok, here's the situation. We're running low on ammo already, the three of us are going out to get more before we run completely dry. Any questions?"

Neither Marine had any.

"Ok, let's go," Pacini said and opened the door.

The three Marines rushed out in a mad dash and immediately little eruptions of sand appeared around their feet as the Iraqi soldiers in the area adjusted their fire directly at them.

From the roof the five Marines who had been left behind to provide cover fire did their best to distract the Iraqis from the three Marines who had left the house. The problem was there were simply too many Iraqis for them to draw the attention of and sand plums continued to kick up around the feet of Pacini, Alverez and Newtson as they ran to the track they had abandoned.

Inside the broken down track, the three of them loaded up with as much ammunition and grenades as they could possibly carry and sprinted back to the house as fast as they could. With the added weight of the ammunition, they didn't cross the distance anywhere near as fast as they had on their way to the track, but, despite the fact that all three of them had had close calls, they made it back without any of them being shot.

As soon as they were back inside the house, which Pacini now thought of as The Alamo, they all took a minute to steady their breathing. Once their breathing was back to normal, Pacini looked at the other two and said, "Take what you'll need."

Since neither Alverez nor Newtson had an M-203 attached to their M-16s, they only took bullets. Once they'd taken enough to last them for a while and returned to their firing positions, Pacini went back up to the roof and ordered the Marines up there to go downstairs and bring everything up. The way he looked at it, he had gone out to the track to get the ammunition, they could bring it up to the roof.

Two hours later, the Marines in The Alamo still had plenty of ammunition left, but apparently some of the Iraqis didn't. From the roof, Pacini saw a group of ten Iraqi soldiers sprint from a house down the street towards the broken down track.

Those slimy bastards are trying to steal our ammo, Pacini thought when he saw what they were up to.

Pacini put a grenade into his M-203 and aimed at the Iraqis, who, thankfully, were grouped too close together for their own safety. He pulled the trigger and the grenade sailed through the air and landed right in the middle of the group. It exploded and sent some of the Iraqis flying through the air. The force of the explosion knocked others to the ground, but most of those of the outer edge of the group were able to remain on their feet.

Even from where he was on the roof it was clear the grenade had killed some of the Iraqis and had left others too badly wounded to even consider trying to make it the rest of the way to the track, or even back to safety. They just lay where they fell, cradling wounded limbs, trying desperately to pull shrapnel out of their bodies. One even looked in shock at his stomach, which had been torn wide open, his internal organs on display for everyone to see. Most of the wounded screamed or cried in pain.

Pacini didn't spend too long watching the wounded because there were others still trying to get to the disabled track. The staff sergeant fired on them with his M-16, dropping most of the Iraqis who had made it past his grenade. The ones he didn't hit got the idea and turned around to run back to where they had come from. None of them made it because Pacini kept firing at them.

He felt bad for the wounded and wanted to put them out of their misery, but in his current situation every bullet and every grenade was simply too valuable to be wasted so he let the wounded suffer until they eventually died.

CHAPTER FORTY-THREE

First Lieutenant Bill Owings, Bravo Company's Executive Officer, had all of his attention focused on firing his M-16 at Iraqi soldiers and hadn't been paying much attention to the effort being made to retrieve the vehicles that had been caught in the sewage trap that had been set by the Iraqis. That was why he jumped when a voice from behind him said, "Lieutenant."

Owings turned around to see who was talking to him and his eyes fell on the sergeant who was in command of the tank retriever that had arrived to free the mired vehicles and help get Bravo Company back on mission.

"What can I do for you, Sergeant?" Owings asked.

"We've got a problem, sir," the sergeant said.

"And what would that be?"

"We went and got ourselves stuck too," the sergeant said, obviously annoyed with himself.

"You what?" Owings asked, though it wasn't so much a question as a statement of disbelief.

"We are stuck now too, sir."

"How did you get yourself stuck? You're supposed to be getting us unstuck."

"Sir, we're just a modified track ourselves. I'm surprised we got as many of your vehicles out as we did before we got stuck," the sergeant said. His voice made it clear he was doing his best not to blow up at the officer.

"So, what are our options, Sergeant?" Owings asked.

"Well, I can call for another retriever... ."

Owings interrupted him by asking, "How long would that take to get here?"

"A couple of hours, sir."

Owings face showed just how little he liked that answer. "Any other options?"

"Only pulling out and leaving what's stuck, stuck, sir."

A few seconds passed as the lieutenant considered the options.

"How many vehicles did you get out, Sergeant?" Owings asked.

The sergeant thought about it for a second. "All but one Hummer and the tracks, sir. I take that back," he corrected himself. "We did get one track out."

When Owings didn't say anything right away, the sergeant asked, "So, do I call for another retriever, sir?"

Shaking his head, Owings replied, "No, no, don't do that. It'll be dark before they could get here and bad things happen in places like this after dark."

"So what do you want me to do, sir?" the sergeant asked.

"Zero out the radios on all vehicles that are still stuck, strip them of anything that we can use, or that the Iraqis can use against us, then blow them up. After all that's done, get everyone loaded up with what we have. We're getting out of here." Owings answered.

"Yes, sir," the sergeant said and set off to do what he had been ordered to do.

CHAPTER FORTY-FOUR

The Marines of Alpha Company had just begun to get over the shock of having one of the tracks from another company show up at their position unexpectedly only to be blown up. As if the day hadn't been hard enough on them already, that experience had shaken most of them to their very core.

When a second track appeared at their position unexpectedly, most of the Alpha Marines prepared themselves for a repeat performance of the disaster they had already witnessed. The Alpha track, in which the dead and wounded had been loaded, had pulled out less than five minutes before the second track's arrival.

Instead of attempting to drive straight through their position, as the first one had, the second track slowed and came to a stop, leaving Iraqis and Americans alike to wonder what was going on.

There might have been a lot of people asking questions, but the appearance of the second track wasn't enough to cause the firefight to lessen any. It continued just as savagely as it had been since the first bullets were fired.

The second track's back hatch opened almost immediately after it came to a stop and the Marines inside poured out to join the fight that Alpha Company was in. They scattered, some positioning themselves alongside Alpha Marines, others finding their own firing positions.

Private Andy Irvin ran to where Private First Class Leonard Karn had positioned himself behind the remains of a burnt out, shot up car. Karn had been exchanging fire with a group of four Iraqis on the entry level of a house and, without any words being exchanged between then, Irvin found a relatively safe place to kneel and fire his weapon.

Several minutes passed before the first Iraqi, who had been popping up in the doorway to fire at them, fell backwards, his chest riddled with bullets from Irvin's M-16, and didn't move again. The others on that floor didn't seem to notice the death of their comrade because the incoming fire didn't lighten for even a moment.

Karn scored a hit, which was more luck than anything else, to the head of a second Iraqi soldier through a window. The enemy soldier fell and didn't return to the fight.

That lucky shot was followed within seconds by one from inside the house that shattered the glass of the car's passenger side-view mirror. The glass flew away from its mounting and a piece sliced across Irvin's cheek. It wasn't a severe cut, but it bled more than a decent amount and would leave Irvin with a nasty scar on his face to remember the Iraqi city of El Jasiph by.

Another several minutes passed without either side scoring another hit. Then, as if they had been hit with a sudden rush of madness, both of the remaining Iraqi soldiers burst forth through the doorway, firing their AK-47s wildly in the general direction of the car that the two Marines had taken cover behind.

Karn, who had been forced to dive to his stomach to avoid the sudden rush of bullets, returned fire from underneath the car. His bullets hit one of the Iraqis in the ankles, turning them to mush, and he fell to the ground in too much pain to even consider firing again. Karn finished him off by firing a controlled burst into the side of the man's torso.

Irvin had repositioned himself towards the front of the car and none of the remaining Iraqi's bullets had come anywhere near him. He let his sight fall on the Iraqi's chest, exhaled and pulled his trigger. His bullets hit the Iraqi right around where his diaphragm was, the force of the bullets impacting him throwing his body backwards as if he was no more than a little girl's doll.

Both men took a moment to make sure none of the four Iraqis were going to move again and then looked at each other. Karn pulled a pack of Marlboro Reds out of his pocket, stuck one in his mouth and offered the pack to Irvin.

"Smoke?" Karn asked.

Taking one, Irvin gave him a grateful smile. "Thanks. I ran out about three hours ago and have been hurting ever since."

"I'm Leonard," Karn introduced himself without shaking hands.

"Andy," Irvin replied.

"Where you from?"

"Charlie," Irvin answered. "What happened to our track?"

"An RPG, I think. It happened so quick I can't be sure."

"What about the guys inside?" Irvin asked, concern for Farrow flooding him immediately.

"Everyone was either killed or wounded. I think more were killed than wounded, but I don't know for sure. Where were they going?"

"Battalion Aid. Everyone was already wounded and we were trying to get them out. Do you know anything about a guy named Burt Farrow?"

Karn shrugged. "Sorry, I don't know who was who, but it was a hell of a mess."

The news caused Irvin to sigh sadly and worry involuntarily. No other words were exchanged between the two men as they rejoined the fight. Irvin didn't trust himself to speak.

Every Charlie Company Marine from track two had a mission in mind as they charged out of the track. For most of them, it had been the same as Irvin's: to join the fight alongside the Marines from Alpha Company. Sergeant Ed Paige, also known as Sergeant Fuckin', had a different mission in mission in mind when he disembarked. His mission had been to find help for Charlie Company, who he was certain was still hanging on by a thread and nothing more. He knew that if at least one of the other companies didn't arrive, and arrive soon, Charlie Company would be completely wiped out.

The second he stepped foot out of track two, Paige sprinted to Alpha Company's command track and ran straight up to Captain Dave Callen.

"Can I help you, Sergeant?" Callen asked once Paige's breathing returned close enough to normal for him to speak.

"Sir, you need to get your company moving over to Charlie's position now," Paige answered, still struggling to speak and trying not to live up to his nickname, a feat which he found very difficult.

"If you haven't been able to tell, Sergeant, we kind of have our hands full where we are," Callen replied.

"But, sir…" Paige began to reply, struggling even harder not to live up to his nickname now that his gut told him that Alpha Company's Commanding Officer would not move to support Charlie Company.

"No, Sergeant, no," Callen interrupted him. "I cannot afford to send anyone to Charlie's aid right now, I need every last man I have right now, including you and each of the men you came in with. Now kindly get your ass out of my track and get back to fighting."

That was the last straw as far as Paige's calm was concerned.

"God-fuckin'-damnit!" he barked. "What kinda fuckin' chicken shit C.O. won't go to help other Marines when they're fuckin' needed. Shithole, we're getting our fuckin' asses handed to us and it wasn't fuckin' supposed to be that way! Your fuckin' ass and Bravo's C.O.'s fuckin' ass was supposed to fuckin' show up and help us out instead of playing your own little fuckin' games in other places! We're getting our fuckin' asses handed to us, asshole!"

Everyone, including Captain Callen, stared in stunned silence as Paige turned around and stormed towards the command track's cargo hatch. No one had ever expected to see a sergeant go off on a company commander that way. It just wasn't how things were done.

Paige made it half way down the cargo hatch's ramp and barked out, "Goddamn fuckin' yellow bastard!" Then he disappeared from their sight.

Once he was gone, the shocked silence hung over the track for a few seconds before Callen turned to one of his privates. "Did I just face the wrath of the infamous Sergeant Fuckin'?" he asked.

"Yes, sir, you did," the private answered.

Furious over Captain Callen's refusal to move in support of Charlie Company, Paige went in search of someone who could override that refusal. His anger burned so brightly he was oblivious to the Iraqi bullets flying within inches, and in some cases centimeters, of him.

Luckily for him, Paige's eyes fell on Lieutenant Colonel Everett, puffing on another one of his cigars. Paige's first instinct was to march right up to his Battalion Commander and rip him a new orifice to defecate out of, but quickly realized that if going off on Callen the way he had hadn't already ended his career, behaving that way to Colonel Everett certainly would. Paige forced himself to calm down as he approached the Battalion Commander.

"Colonel Everett, sir," Paige said crisply and professionally. The way he should.

"What is it, Sergeant?" Everett asked, clearly annoyed by the interruption.

"Sir, Charlie Company is being wiped out. We need help and we need it now, sir," Paige said, making every effort not to live up to his nickname.

"Is it really that bad over there?" Everett asked.

"Yes, sir. Sir, if we don't get help soon, there will be nothing left of Charlie Company but a bunch of Marines in body bags."

Paige's sense of urgency was evident in his voice, his body language and his eyes.

"There's nothing I can do at the moment, Sergeant," Everett said and Paige felt himself on the verge of blowing up again. How couldn't he? Charlie Company, his friends, was up against the ropes and wouldn't be able to hold on much longer on their own. "However, as soon as Alpha's tanks show back up, I'll send two of them to reinforce Charlie until we can get this mess here cleaned up."

Paige felt a great burden lifted off of his shoulders, Charlie was finally going to get the help they needed so badly.

"Sir, if I can ask, how long will it be before Alpha will be able to move in support of Charlie?"

"Not long, Sergeant. I'm done farting around here and, civilians or no civilians, as soon as we can get through on the Battalion TAC, I'm calling in an airstrike to level these buildings. It's nice to protect civilians when we can, but not at the cost of my Marines. It might cost me my career, but the gloves are coming off and we're going to wrap this up," Everett answered.

"Thank you, sir, thank you," Paige said.

"Now, Sergeant, please put your weapon to use and kill some fucking Hajis."

Paige smiled. "Yes, sir, right away, sir!" he said and ran off, much happier than he had been when he had first arrived.

CHAPTER FORTY-FIVE

The going wasn't easy for Corporal Marcus Andrews and Private First Class Lee Newickki as they made their way through the mean streets of El Jasiph. Being unable to see, Newickki was a nervous wreck. He could hear the bullets flying through the air around them, many of which came way too close for comfort, but he couldn't see where they were coming from or who was firing them. He didn't like it in the least.

For Andrews, who still didn't realize that he had been shot in the back, things weren't quite as nerve wracking because he could see who was shooting at them. As they moved through the unpaved streets, Andrews knew he couldn't exchange bullets with every Iraqi who shot at them as he didn't have enough ammunition and he was only one man. If the person shooting at them was outside, on the first floor of a house or in a courtyard for example, he would engage them—he had to, there wasn't any way around it. Andrews also engaged those he didn't think he could get both of them past safely. If he thought he could, Andrews tried to sneak both of them past without being noticed. Sometimes he succeeded, sometimes he ended up trading bullets with Iraqi soldiers anyway. He didn't know how many he killed on his trek through El Jasiph, but it was quite a few.

From the roof of the Alamo, Corporal Stan Fackler took careful aim at the numerous Iraqi soldiers, who made the mistake of popping their heads up when he was looking in their direction. The problem with urban warfare is that there are always plenty of civilians around to show up in the wrong place at the wrong time so he didn't pull his trigger until he saw a weapon. The last thing he wanted was to have to go through the rest of his life carrying with him mental pictures of the faces of innocent women and children who were accidentally killed by his weapon.

The sight of two people making their way slowly, almost too slowly, through the streets caught Fackler's eye. His first instinct was to point his weapon at them, but something, he couldn't say what, caused him not to.

He watched the two people for a few minutes. Slowly, his conscious mind realized that he was looking at two Marines, not two Iraqis as he had originally believed. He could tell that one of them was badly wounded.

"Staff Sergeant!" he barked out. "We've got two of our guys down on the street!"

"What?" Pacini called back.

"We've got Marines coming in on foot!"

Pacini dashed across the roof. Once he was there, Fackler pointed to where Andrews and Newickki were on their desperate flight through the city and Pacini's eyes followed.

Christ on a cracker! Those guys are sitting ducks out there! Pacini thought when he saw them.

Ignoring the risk to himself, Pacini stood on the wall that extended above the roof. Waving his hands, Pacini shouted, "Marines! Hey, Marines! Over here! Come this way!"

There wasn't any way for him to be sure, but Pacini thought he saw one of the Marines look at him. Time would tell. There were still plenty of bullets flying through the air between the Marines on the street and the Alamo, but Pacini hoped they would make it to them.

"Marines!" Newickki and Andrews heard someone yell. "Hey, Marines! Over here! Come this way!"

Andrews's eyes immediately began scanning the cityscape, especially the rooftops.

"Who was that?" Newickki asked, his head turned in the general direction that the voice had come from, his worthless eyes searching for the source even though they were no longer capable of sight.

"No idea, man, no idea, but I'm looking," Andrews replied as he continued to take in everything around him.

After a few seconds Andrews saw Pacini standing on the wall of a house waving his arms. He wanted to wave his arms in reply so the Marine calling to him would know he had heard him. He couldn't though, Andrews didn't know why, but he couldn't raise his arms any higher than chest level. He still hadn't realized that he had been shot in the back.

Even though Andrews couldn't wave back, his heart jumped with joy at the sight of the other Marines. He and Newickki were not out here on the streets of El Jasiph on their own.

"Right now, that's more of a beautiful sight than anything I've ever seen in a strip club," Andrews said to himself.

"What was that?" Newickki asked.

"New Hickie, we've got Marines! We're not alone out here!"

"Well what the hell are you waiting for? Get us to them."

Andrews took off. While he was still being cautious and shooting at Iraqi soldiers as he walked, there was a new vigor in his steps. Help wasn't as far away as he had thought.

Newickki, while still scared because he couldn't see anything that was going on around him and in the worst pain he had ever experienced in his life, was hopeful. Ever since the track he had been riding in was destroyed, with him in it, he had worn a sense of fatalistic pessimism like a tight fitting outfit. Now that he knew he and Andrews weren't out there on their own, that there were other Marines fighting where they were, his pessimism quickly faded away and a feeling of hope replaced it. Maybe, just maybe, he would make it through this and then maybe the doctors could give him his sight back.

The two Marines fought their way to the Alamo, where Alverez quickly opened the door to let them in.

"Welcome to the fabulous Raghead Resort and Casino. I'm Lance Corporal Newtson…Marcus, is that you?" Newtson said as Andrews and Newickki walked through the door.

"Tim, oh shit, man, I didn't think I'd ever see you, or anyone else, again," Andrews said.

"We thought you were dead, we saw your track get blown up."

"We can talk about that later. Newickki here is hurt bad," Andrews said and then fell to the floor, face first, hitting the ground hard.

"Shit, you are too. Manny, take New Hickie to the causality collection point and then come back here. He's shot in the back, real close to his spine. We should carry him."

Alverez didn't say anything in reply to Newtson. Instead he grabbed hold of Newickki's arm and said, "Follow me."

On the roof, Pacini settled back into his original position. He hadn't been able to tell who the two new arrivals were, but he was glad they had made it.

Two old and very poorly maintained motorcycles drove right past the Alamo. Each motorcycle had a driver and a passenger. The drivers kept their eyes on where they were going and the passengers fired AK-47s at the Alamo.

Pacini opened fire on one of the motorcycles, hitting the driver in the head. After dying, the driver jerked the bike sharply to the right, causing it to fall over. The passenger tried to escape, but Pacini stitched American-made bullets up his back from his waist to his neck.

The second motorcycle was gone before Pacini could aim at it. Surprising the staff sergeant, it came back seconds later, firing at the Alamo as it drove by. Pacini fired and hit the bike's gas tank, which promptly blew up and turned both riders into human torches.

After removing all four Iraqis from this world, Pacini dropped down behind the wall, ejected his empty magazine and replaced it with a full one.

Hopefully the next Marines to come along will be coming to rescue us, he thought as he rose back up and started firing at Iraqis in a building directly across the street from the Alamo.

CHAPTER FORTY-SIX

Alpha Company's situation grew worse by the minute. A few of them had been killed, but their number of injured grew rapidly. More and more of their vehicles were either on fire or were now burnt out wrecks. The Iraqi soldiers may have been bad shots when the Marines first arrived in El Jasiph, but in the hours since they seemed to have gotten better at shooting Marines who really didn't have anywhere to go and they'd grown quite proficient at hitting vehicles that were sitting still. Despite everything they had going against them, Alpha Company was still in much better shape than Charlie Company.

With a cigar in his mouth, the Battalion Commander, Lieutenant Colonel Jim Everett sat on the street with his back against his command Humvee and studied a map of the city. Even though Alpha Company had fought well and done the Corps proud, he knew it was just a matter of time until the tide turned on them too greatly to recover from. It was a simple matter of numbers—the Iraqis had them and the Marines didn't.

Although the similarity had not struck any of them yet, many of the Marines who both saw Colonel Everett at that point in time and who survived the day would be struck by how much their Battalion Commander resembled the legendary Marine, Lieutenant General Lewis Burwell "Chesty" Puller. Chesty Puller was the most highly decorated Marine ever and saw combat against guerrillas in Haiti and Nicaragua as well as having been in some of the bloodiest battles of World War II and Korea. It was a very favorable comparison and it would have made Everett proud, as both a man and a Marine officer, to know his men held that kind of image of him.

The realization of just how desperate their situation would eventually become, and the news about how desperate Charlie Company's situation had already become, prompted Everett to pull out his map and plot the quickest route to Charlie Company's position. He realized that the path laid out for him in the planning of the day's operation probably would not be the fastest route and he wanted to get Alpha and Charlie Companies together as soon as humanly possible. Together they would stand a much greater chance at surviving and pulling out a success where the possibility of defeat was growing ever more likely.

There was only one hitch in his plan, he couldn't move Alpha Company forward until this area of the city was secure. If he did, his Marines would just have to come back and fight here again, which they would probably win, but it was an unnecessary step as long as he could get through on the Battalion TAC to call for air support. His problem there was that he couldn't get through.

As if he had been reading his Commanding Officer's thoughts, Sergeant Major Art Jamison came running up and dropped down next to him.

"News from Alpha's Command track, sir," Jamison said.

"What is it?"

"We got through on the Battalion TAC. We've got Cobras inbound to our position to lend us a helping hand," Jamison answered with a smile on his face.

Everett gave him a questioning look. "Did you say Cobras, as in plural, Sergeant Major?"

Jamison's smile grew even wider. "Yes, sir, as in more than one."

"How many?"

Jamison shrugged. "No idea, sir. They just said they had a few that had returned for refueling and rearming and they'd be on their way here as soon as they were airborne. There will be a Sea Knight on their heels to take the dead and wounded off of our hands too."

"Well that's some good news at least. Were they able to give you any word on Bravo Company?" Everett asked.

Sergeant Major Jamison shook his head. "Nothing, sir."

Everett nodded his head and turned his head to the sky, his eyes eagerly searching for the distant shapes of approaching helicopters.

Twenty minutes later, just the opening and closing of an eye compared to how long the Marines of Alpha Company had been holding their position, six AH1Z Super Cobra helicopters appeared and began firing at the buildings around the Marines. The Marines couldn't help but cheer as the Cobras opened up with their twenty millimeter Gatling guns and fired their Hellfire missiles, Sidewinder missiles and rocket pods into the buildings.

The bullets from the helicopters' twenty-millimeter guns turned any Iraqi unfortunate enough to be caught by them into human burger while the missiles and rockets reduced the buildings they were in to rubble.

Dust caused by the collapsing buildings filled the Marines' lungs and caused them to cough, but not even that could reduce the jubilant feeling they had. The helicopter assault only lasted a few minutes and then they moved on to different parts of the city. After they left, the devastation the helicopters had brought to Alpha Company's area of El Jasiph was clearly evident too. The best result, however, was the fact that no one fired at them again in the absence of the Cobras.

Within minutes of the Cobras' departure, a single, duel rotor CH-46 Sea Knight helicopter landed on the street. The dead and wounded were loaded in less than five minutes and the Sea Knight took off again, heading for the battalion aid station.

The third track of Charlie Company's medevac convoy had already arrived and kept driving on through Alpha Company's position and on to the battalion aid station on the other side of the Euphrates River, just as the ill-fated first track had hoped to do.

CHAPTER FORTY-SEVEN

Tensions were high inside the Alamo, but the Marines trapped there didn't allow it to make them start turning on each other. They were under constant fire, which ratcheted up their stress levels to a degree none of them had ever thought possible, they were hot, they were sweaty, and they were miserable. The sounds of pain coming from the casualty collection point were another constant that did nothing to reduce their stress levels. They all wanted to do something to help relieve their brother Marines' suffering, but there simply wasn't anything they could do. They were helpless, and being helpless was a feeling none of them liked very much.

As he lay on the ground in the casualty collection point, Private Lee Newickki felt a vibration on the ground, which was followed shortly by a rumbling sound. His lack of vision caused every sound Newickki heard to disturb him, but he found the rumbling and the vibrations particularly disconcerting.

"Hey!" he called out. "Hey! Is that a thunderstorm moving in or something?"

Private Alverez and Lance Corporal Newtson were too busy fighting to hear him. The other unwounded Marines were on the roof so they couldn't hear him either. Those in the casualty collection point with him were too wrapped up in their own personal misery and suffering to worry about what he was saying. So no one answered Newickki.

A few minutes went by and nobody answered him, so Newickki yelled out, "Hey you motherfuckers! Answer me, goddamnit! What's that rumbling sound?"

When no one answered him that time, Newickki didn't bother asking again. He knew the unwounded guys in the Alamo were busy trying to keep them all alive and probably had other things on their minds than answering his questions. They probably hadn't even heard him.

Newickki wasn't the only one who had heard the rumbling sound. Up on the roof, Pacini also heard it and looked around.

"Hell yes!" he shouted when he saw the two tanks from Alpha Company on their way to Charlie Company's position. "We've got tanks! You hear me boys? We've got fucking tanks!"

Gotta get their attention, Pacini thought. *Gotta let them know we're here. How?*

"Flares! Did anyone grab any flares from the track?" he shouted loud enough for all of the men on the roof to hear him.

"I saw some somewhere. Hold up!" one of the Marines answered.

The Marine jumped up from his firing position and ran to where they had piled a bunch of gear together. "Got 'em," he said.

"Bring them, quick! These tanks are our way outta here!" Pacini replied.

The Marine ran over to Pacini and handed him three red flares.

Looking at the young private, Pacini said, "Thanks, now get back to firing."

He turned his attention to the approaching tanks. Pacini lit the first flare and dropped it from the roof, hoping it would catch the attention of at least one of the tankers. The thought that the one flare would have been enough to tell those in the tanks where they were hadn't crossed Pacini's mind. All he hoped was that they'd see it and start paying attention.

Once the tanks were close enough to see where the second flare fell from, Pacini lit it and dropped it to the sand covered street below. After it was dropped, he had to force himself to be patient and not to drop his last flare until after he was certain the tanks were close enough to see him.

A few minutes went by and the two tanks came close enough that Pacini was positive that, as long as they were looking for where the other two flares had come from, they couldn't miss the third one. He lit the flare, held it over the side of the wall and let it drop.

Within seconds the tanks rolled up outside of the Alamo and came to a stop.

"Hell fuckin' yeah! They stopped! Hell fuckin' yeah!" Pacini yelled as he ran down the stairs.

Pacini ran from the stairs to the door and pulled it open as fast as he could get his arms to work. He opened it just as a major walked up to it.

The major smiled and asked in a thick Cajun accent, "Ya'll mind if a feller comes on in? The natives seem ta be a might restless out here today."

Dumbfounding Pacini, the major just stood there outside the door as if he were back home and had just walked up to a neighbor's house on a leisurely Sunday afternoon. He didn't seem to notice the chips of cinder block were being knocked off the wall by Iraqi bullets.

"Come in! Damn, man, come on in before you get your ass shot off!" Pacini yelled, unable to believe the officer was just standing there, waiting to be invited in. It made Pacini, a lifelong horror fan, think of how, according to some authors, vampires were incapable of entering a house without being invited in first.

The major stepped inside and Alverez slammed the door closed behind him.

"Major Sean Raynes," Major Raynes said, holding out his hand to Pacini, expecting him to shake hands.

This guy is unlike any officer I've ever met. He must have been a Naval ROTC student at Redneckville University or something, Pacini thought. He shook Raynes's hand. "Staff Sergeant Nick Pacini, sir." Little did Pacini know that Major Sean Raynes graduated top of his class from Annapolis and was known as a real hard ass by those under his command. His laid back, easy going demeanor was only an attempt at helping the enlisted men in the Alamo, who were all clearly, and justifiably, stressed, relax.

"Nice ta meet ya, Staff Sergeant," Raynes said. "Now, here's what I got in my noggin. I see you've got wounded here and I don't have any room in my tanks for hitch hikers. I say we put the wounded on the outside of the tanks and then the rest of you patrol on out of here, on foot, right next to us."

"I like the idea of you taking the wounded outta here, sir, but I'm not big on the idea on the rest of us trying to walk out of here. Things are nuts out there and there are a lot of miles between us and safety," Pacini replied.

"What're you thinkin' then?" Raynes asked.

"Well, sir, if you don't have room to give us all a ride, I think it would be best if you just take the wounded with you and then whenever you get to where you're going you can let them know where we are. Someone else can come back to give us a lift. No offense, sir, but Haji is everywhere out there, inside houses, on rooftops, in alleys, behind just about every wall. I think the rest of us would be a lot better off if we stay where we are until someone else can come back for us, instead of us trying to walk out and getting shot up on the way."

Raynes nodded. "Makes sense ta me. My Marines are all busy keeping Haji too busy to shoot at us too much. Your men will have ta help me load up your wounded."

"Sounds good to me, sir," Pacini replied.

With their numbers as depleted as they were, Pacini had to have every Marine in the Alamo help load the wounded onto the exterior of the two tanks. Within seconds of the last wounded Marine being loaded up, the tanks were once again rolling towards Charlie Company's position.

By the time the tanks were gone, all the Marines from the Alamo had resumed their previous positions and had reengaged the Iraqi soldiers who seemed to be coming from everywhere.

CHAPTER FORTY-EIGHT

Alpha Company had had it a lot easier ever since the Cobras came in hot and heavy. An occasional Iraqi would take a potshot at one of the Marines, but the incoming enemy fire wasn't anything like it had been. The Marines were now able to walk around freely without being torn to pieces.

With his cigar still clenched between his teeth, Lieutenant Colonel Jim Everett stepped into Alpha Company's command track. "Captain Callen," he said.

"Yes, sir," Callen replied.

"Are all of your Marines accounted for?"

"Yes, sir."

"How many prisoners did we pick up?" Everett asked.

"Sixteen, sir," Callen answered.

"Ok, get your men and the prisoners loaded up. We're going to go give Charlie Company a hand. I want us on the move five minutes ago. Understand me, Captain?"

"Yes, sir."

Callen looked at Alpha Company's First Sergeant, Hector Urness. "First Sergeant, go and find each of the Platoon Commanders and tell them it's time to get rolling. They are to get everyone loaded up and ready to roll with all possible speed."

"Right away, sir," Urness replied and disappeared from the track.

In less than twenty minutes, Alpha Company was once again on the move. Since they had lost vehicles, gained prisoners and gained Marines from Charlie Company, conditions in the tracks were a lot more cramped than they had been on the way into El Jasiph. The addition of one of Charlie's tracks helped out some, but conditions were still tight.

The fact that none of the men had had a shower in several days combined with the stress of the last several hours and the heat of the

Iraqi desert had cranked up their body odor to a level it really was never meant to be cranked up to. The stench of unwashed bodies along with the cramped conditions in the AAVs combined to make the Marines miserable during the rest of their journey through El Jasiph.

The enlisted men hadn't been made privy to the route the company would take through the city, but they felt the motion of the track turning so they knew they had left the street where most of them had experienced combat for the first time.

Ten minutes into their journey, bullets started slamming into the side of the track Lance Corporal Barry Queen rode in. Many of those bullets broke through the track's hull and ricocheted around.

One of them hit Private First Class Chris Martinez in his right knee and he collapsed to the floor in a near fetal position, screaming in pain. Queen rushed over to help him, quickly tore Martinez's pant leg open so that he could see the injury and used both hands to apply pressure to both the entry and exit wounds. "Get me a first aid kit!" he yelled.

Someone, Queen didn't see who, dropped a first aid kit on Martinez's stomach. Queen let go of the wounds long enough to open it, get out what he needed and set to work stopping Martinez's bleeding.

Lance Corporal Victor Lauralwood still manned the Mark 19 grenade launcher of one of the CAAT Humvees, just as he had been since well before the Marines entered the city of El Jasiph.

He saw an Iraqi soldier start firing on the track carrying Queen and Martinez from the roof of a house with an old M-60. Lauralwood squeezed the trigger of his Mark 19 and sent ten grenades sailing through the air to the Iraqi soldier's position.

All ten grenades landed and all ten exploded. Lauralwood would never know if he killed the Iraqi or not, but he could take pleasure in knowing his quick reaction probably saved the life of at least one Marine because the enemy soldier didn't fire at them again.

CHAPTER FORTY-NINE

Incoming fire at the Bravo Company Marines, who were waiting for those with the vehicles that were mired in sewage, was constant, but it was only heavy every now and then. The biggest threat came from a heavy machine gun that, despite their best efforts, they couldn't get to shut up.

"Jesus H. Christ!" Captain Dan Earl, Bravo Company's Commanding Officer yelled. "When will that guy either die or catch on to the fact that we don't want him shooting at us!" He sat on the ground with his back to one of the company's Humvees.

"Seems to be a thickheaded one to me, sir," First Sergeant Warren Backer replied.

"How many of our guys has he hit now?"

"No clue, sir. I could find out if you'd like," Backer answered.

"No, that won't be necessary. What haven't we tried? What else can we do to shut that gun up?"

"Snipers, sir?" Backer suggested.

Earl looked at him. "You think so? That's Haji held territory. A lot of ragheads over there. You think one could reasonably be expected to get into a position to take him out without being killed himself?"

Backer shrugged. "Couldn't hurt to try, sir."

Earl looked away from and him seemed to think about it for a few seconds, then he looked back at his First Sergeant and said, "Find one and send him."

"Just one, sir? We have four," Backer replied.

"Yeah, just one. We might need the others to silence that machine gun. I'm going to send one at a time and hold the rest in reserve. If the first doesn't make it to a spot to take Haji out, maybe the second will, but that second sniper won't get the chance if I send all four at once and they all get killed trying to get to him."

"Makes sense to me, sir. I'll get right on it," Backer said, stood up and ran away from Earl, hunched over to make as small a target of himself as possible.

As he ran, Backer attracted an increased amount of enemy fire, but, fortunately, the Iraqi firing the heavy machine gun didn't seem to take any notice of him. The first sniper he came to was Corporal Alan Wakeland, who was sending Iraqi soldier after Iraqi soldier to meet Allah face to face, one shot at a time.

"Wakeland!" Backer yelled to be heard over the sound of gunfire.

The sniper quit shooting. "Yes, First Sergeant!"

"I want you to go and take out that machine gun before he takes all of us out."

Wakeland looked over to where the muzzle flashes from the machine gun were coming from, clearly deciding where he would need to be in order to take that shot.

Won't be easy without a desert Ghillie suit, but I've done worse on training exercises, Wakeland thought. "Too easy, First Sergeant, too easy."

Nothing else needed to be said. Wakeland slung his M40-A3 over his back, pulled out the nine-millimeter Beretta he carried when it wasn't practical for him to have his sniper rifle at the ready and took off.

A little more than twenty minutes later, Wakeland found the spot he wanted to take his shot at the machine gunner from, on the roof of a building that was three buildings down from the Iraqi's position. He slid his Beretta back into its holster, took his rifle off his back and went to work placing the Iraqi soldier's head in his crosshairs.

Regardless of what the movies said, head shots weren't the optimal shot a sniper wanted to take, but, in this case, it was the only shot he had. The countless hours of training Wakeland had been through prevented him from even being tempted to rush the kill shot. He might only get one shot, so he had to make it count. He steadied his breathing, waited for just the right moment, and gently squeezed the trigger when it came.

His bullet flew through the air and struck the Iraqi in his right ear canal. It may have made a big gory mess, in fact it probably did, but all Wakeland saw was the Iraqi slump over to his left and remain still. The machine gun that had been making the Bravo Marines' lives a living hell ever since it started barking hot lead towards them fell silent.

With that Iraqi's death, the amount of enemy fire coming down on the Marines in that portion of Bravo Company decreased dramatically. Wakeland took a minute to study his surroundings to see if there were any targets of opportunity that were ripe for the picking. He didn't see any.

It can't be too much longer until the others catch up to us, the sniper thought. *When they get here it'll be time to go and help Charlie out, so I can't afford to spend too much time farting around out here.*

As much as he wanted to get back to the safety the other Marines provided, the last thing Corporal Wakeland wanted was to run into friends of the man he had just killed. The desire not to run into any hostiles while he was by himself caused Wakeland to move slowly and cautiously as he made his way back to the portion of Bravo Company that he had moved ahead with.

CHAPTER FIFTY

Sergeant Rick Neighbors had worked one of the CAAT Humvee's fifty-calibers as hard as he could since Charlie Company first rolled into the city of El Jasiph. It had been hours, and his ammunition had held out. Neighbors had been so focused on what he was doing it hadn't even dawned on him that he could be running low on ammunition, or even that he could run completely out.

His weapon ran dry and Neighbors dropped down into the Humvee to grab more ammunition and reload, which he had done repeatedly during the firefight. His eyes grew wide in surprise when he couldn't find anything besides empty shell casings.

"Shit!" he yelled when the realization that he didn't have anything to reload his weapon with hit him.

Neighbors jumped out of the Humvee and started running to the nearest track. Once again he was focused on the task at hand, which, unfortunately, had the same effect on him that blinders had on race horses, and he tripped over the legs of Corporal Chris Jabcon, who was lying flat on his stomach as he fired back at the city. He didn't land gracefully.

Annoyed worse than he could remember ever being, Neighbors spit out a mouthful of sand and looked back to see what he had tripped over. "Son of a bitch!"

When he saw it was another Marine he tripped over, Neighbors ordered, "Come with me. I could use some help."

"Right behind you," Jabcon replied, jumped to his feet, and followed him without having any idea of what they were up to.

As soon as they were inside the track, Neighbors grabbed a box of fifty-caliber ammunition and handed it to Jabcon. He then grabbed a box for himself.

"Take it to my Humvee," Neighbors ordered.

Nothing else was said as the two Marines ran from the track and covered the distance between it and Neighbors's Humvee. Once they made it, they both loaded their ammunition boxes into the backseat and Neighbors went to work reloading his weapon while Jabcon started firing his M-16 from the cover the Humvee provided.

Neither of them had been hit, but they could hear the cries and screams of those who had been. They both knew Charlie Company was on the razor's edge of being obliterated entirely.

"Shit, Sergeant, if the others don't show up soon, we're all well and truly fucked. You know that don't you?" Jabcon said looking up at Neighbors as he reloaded his M-16.

Neighbors, who hadn't heard what he said, quit firing his weapon and asked, "What?"

"I said, if we don't get some help soon, we are well and truly fucked!"

Neighbors looked around. He saw the torn, broken and bloodied bodies of U.S. Marines lying on the Iraqi sand, some of whom were being attended to by their fellow Marines since Trevino had gone with the medevac convoy. Sadly, others were lying on the ground and screaming or crying in pain while they waited for someone to come and try to help them. With the battle raging around them, not every Marine could stop fighting to do what they could for the wounded. He saw the burnt ruins of some of their vehicles. He hadn't really taken the time to look around and take in his surroundings up until then, but everything he saw pointed to the near inevitability of Jabcon's conclusion.

"You know, I didn't realize it until just now, but you're right. We're pretty bad off aren't we?"

Before Jabcon could reply, the sound of approaching tanks hit their ears and cold knots developed in the pits of their stomachs.

"I hope those are ours," Jabcon said, "because if they're Haji, we're dead."

Neighbors turned his fifty-caliber just a bit to face the sound of the oncoming tanks. "That might be, but I don't plan on letting them kill me without at least making them remember that I was here."

Several tense minutes passed and neither of them spoke. They barely even breathed. When Alpha Company's two M1-A1s appeared, they both breathed a sigh of relief. The stress of the moment had been weighing on them like a semi-trailer, but that stress disappeared instantly.

"Hell yeah! You see that, Sergeant? They're ours! They're Abrams! Hell yeah!" Jabcon yelled in joy.

Neighbors didn't say anything in reply. He just expressed his relief by smiling and chuckling lightly.

The two tanks instantly drew Iraqi fire. They drove up to several of the Marines from Charlie Company and came to a stop. From where they were, neither Neighbors nor Jabcon could hear what was said, but they saw the hatch of one of the tanks open up and someone rise up out of it. The man in the tank said something to the Marines on the ground and they jumped to work unloading the Marines who were on the exterior of the tank. Neighbors could only assume the Marines being offloaded were wounded.

Once the wounded were off, the two tanks drove away from each other, one to each end of Charlie Company's position, and began firing into El Jasiph. The booming of their one hundred and twenty millimeter main guns was deafening, but the Charlie Company Marines were more than happy to have the loud booming on their side.

A building in El Jasiph collapsed into a pile of rubble with each shell the Abrams fired. After each tank fired five shots each and ten buildings had fallen into rubble and a choking cloud of dust, the incoming Iraqi fire died out.

An excited cheer from the Marines met the lack of incoming fire, but their joy was short lived. Shells from an Iraqi artillery position started falling on Charlie Company's position, trying to take out the two tanks that had come to their assistance, and sent the Marines scrambling for cover.

High above Charlie Company's position, the same A-10 that had mistaken Charlie Company's AAVs for Iraqi tanks saw the muzzle flash of the artillery rounds being fired. The pilot put his aircraft into a dive, leveled out and began his attack run.

The second he was in position, the pilot released his bombs. They fell free of the Warthog and seconds later frightful explosions appeared on the ground. Once the explosions ended, the artillery position didn't fire again. The bombs had done their job.

CHAPTER FIFTY-ONE

The *whump, whump, whump* of the dual rotors of a CH-46 Sea Knight broke the quiet that had fallen over Charlie Company's position since the Iraqi guns had fallen silent. The helicopter landed and Marines began loading their dead and wounded friends onboard while others climbed off.

Private Andy Irvin was one who had stepped off. When the fighting at Alpha Company's position had finally broken, he had been part of a small group of the Charlie Marines Lieutenant Colonel Everett had given permission to go to the battalion aid station to visit their wounded. The Battalion Commander knew they were all anxious to find out about their buddies and had let a small group of them go to the battalion aid station to check on them so they would be able to bring word back to the rest of the company on how everyone was.

Now Irvin was back, dazed. The fact that his "brother from another mother", Burt Farrow, was dead hadn't thoroughly sunk in yet. He kept hoping it would all turn out to be a case of mistaken identity and Burt would be waiting for him when he arrived back at Charlie Company's position.

Deep down, though, Irvin knew that wouldn't be the case. He had seen his friend's remains. There hadn't been much left, but between Farrow's dog tags and the distinctive tattoo of a naked woman riding a demonic looking bulldog, or Devil Dog, with USMC scripted below it, Irvin knew the truth, at least on an intellectual level. The fact that Burt Farrow had died at the age of eighteen didn't seem real yet—it couldn't be true.

As soon as the dead and wounded were loaded on the helicopter, it lifted back into the air and took off on its return trip to the battalion aid station.

Major Raynes walked from his tank to Charlie Company's command track. He looked at Captain Aber. "We've got some Marines stuck in a house and under heavy fire. Can you spare a couple of Humvees to go and get them?"

Aber chuckled. "If you'd asked me that question not all that long ago I would have thought today had been too much for you and had you held for a Section Eight evaluation. But, thanks to you guys, I seem to have more manpower than I need. Since you know where they are, would you mind accompanying my First Sergeant to brief the drivers, sir?"

"It would be my pleasure," Raynes replied.

"First Sergeant," Aber called out.

Raab, who had been standing outside the track, was in front of him in seconds. "Yes, sir," he reported.

"I want you to go with Major Raynes here and find five Humvees to go and pull out Marines who are trapped in the city," Aber said.

"Yes, sir," Raab replied.

"Oh, and Captain," Raynes cut in.

"Yes, sir?" Aber asked.

"I would suggest sending your CAAT Humvees with the TOW and the Mark 19. I don't know what's going on now, but when I was there, there was some pretty nasty fighting."

"Yes, sir. I can afford both of them now," Aber said.

"Good. See you later, Captain," Raynes said and left the track with Raab right behind him.

Thirty minutes later, the convoy of Humvees appeared on the street Major Raynes had identified and they immediately came under heavy fire. The driver of the CAAT TOW Humvee had been shot in the throat and died, so Raab sat in that vehicle's driver's seat.

Before they left, Raab had issued orders to the three fifty-caliber gunners, the TOW Operator and the Mark-19 Operator to destroy anything that fired at them. They were not to wait for a fire order from him. Within seconds of the Iraqis opening up on them, the convoy returned fire and wiped out several Iraqi positions along the street on their way to the Alamo.

By the time the convoy arrived at the Alamo, the street had fallen eerily silent and they found the Marines that had taken cover inside standing outside waiting for them. The convoy pulled up in front of the Alamo and came to a stop.

Raab climbed out. "Someone call for taxi service?"

"First Sergeant, I thought I'd never be so happy to see you," Pacini said.

"Pacini?" Raab asked.

"Yes, First Sergeant."

"When I heard there were Marines trapped out here, I had no idea they were our Marines. Get your men loaded up so we can get out of here before Haji decides to go for another try at us," Raab said.

"Yes, First Sergeant," Pacini replied.

He turned to the Marines who had been in the Alamo with him. "You heard the First Sergeant. Get your asses in those Humvees, now."

They didn't need to be told twice.

CHAPTER FIFTY-TWO

As darkness met the Marines who had spent the day fighting in El Jasiph, all three companies had rejoined each other and had set up their new camp. The day had been long and grueling and most of them found it impossible to fully relax as they had been able to back at Camp Nicholas. Had it only been days since the Battalion pulled out of Camp Nicholas to blaze the trail into Iraq? It seemed more like years had passed. Their work in El Jasiph and the area immediately surrounding the city was far from over, but most of them hoped the really heavy fighting was a thing of the past. Lieutenant Colonel Everett didn't want to risk Iraqi reinforcements being brought into El Jasiph during the night so he ordered roadblocks placed at all three entrances into El Jasiph. No one would be entering or leaving the city that night, and possibly even longer.

None of the Marines had had much sleep since the night before they left Camp Nicholas and not too many of them would catch up on their sleep that night either. American artillery and Cobra helicopters laid waste to positions the Iraqis were setting up to continue the fight the next day. The sound of explosions and guns being fired kept most of the Marines awake and on edge.

Private Manny Alverez sat cross legged on the sand, looking into the fire. The expression on his face showed that he was anywhere but there at that point in time. He was lost in his thoughts and worries. He had not seen or heard anything about Andy Irvin or Burt Farrow, the two people who meant the most to him in the entire country of Iraq. Next to his immediate family, they meant more to him than anyone else alive, and he didn't know where they were, how they were doing, or even if they had been two of the many Marines lost that day.

Lost in his own little world as he was, Alverez didn't hear the footsteps of someone approaching behind him and jumped when he felt a hand gently touch his shoulder. Alverez turned and saw Irvin standing there looking down at him sadly. "Andy! Damn, brother man, I thought you'd gone off and died on me or something," Alverez said, his voice making his happiness clear.

Irvin sat down next to him. He couldn't bring himself to say anything and that told Alverez something had gone really wrong. They were both there, physically whole, so the bad news had to be about Burt Farrow.

"Burt?" Alverez asked.
Irvin just looked at the ground and shook his head.

"Andy, tell me, man. What is it? How's Burt?"

This time Irvin looked up at him. Tears were streaming down his face. The sight told Alverez all he needed to know.

"He's….he's dead?" Alverez asked as he felt his eyes beginning to well up with tears of his own.

As much as Irvin wanted to answer Alverez, to talk with him about their friend's death, he just couldn't. The words failed him. Instead of saying anything, Irvin just nodded his head sadly.

"Oh, fuck, man, and here I always thought the three of us would live to be old men, telling our grandkids about when we fought in Iraq together. I… I never thought one of us could actually die. Oh, fuck, man, he was only eighteen, just like us. He can't be dead… not Burt," Manny said, hoping Irvin would start laughing and tell him the whole thing had been nothing but a sadistic joke the two of them had cooked up. But he knew that wouldn't happen. Burt Farrow was actually dead.

Irvin looked at the ground and said, "Sorry, brother." It was all he trusted himself to say.   The two of them just looked at each other silently for a few seconds and then embraced each other tightly. They wept without shame. Modern society tells us it is unacceptable, and even shameful, for men, especially warriors, to cry. After all, no one ever saw John Wayne or Clint Eastwood cry in their movies. Alverez and Irvin now knew the truth: despite what society said, they now understood there is no shame in mourning the loss of a comrade in arms, the loss of a brother. They were both combat veterans at an age where most people are thinking about going to college or working a minimum wage job. They were past caring what everyone else might consider unmanly. They knew they were men. When they separated, Alverez looked at Irvin with tears shimmering in his eyes. "Spic?" he asked, his voice making it clear he was reaching out desperately for the closeness the three of them once held. He needed something familiar.

Irvin smiled sadly. The whole time he had known Farrow and Alverez, he really had wanted them to reach out to him with that little game they played. Now Alverez was, and it made him feel good.

"Yeah, Nigger," Irvin replied.

"Where do you think Burt is now? I mean...do you think he's in Heaven?"

"Hell yeah he is, bro," Irvin answered. "He's up there in Heaven's tittie bar right now, where the long legged beauties are nothing short of womanly perfection and your beer never runs out or gets warm."

Despite his sadness, Alverez chuckled. "And he's looking down at us saying 'Why are you two leaving me hanging? I don't know if I can handle all these beauties on my own. I need my bros.'"

Irvin rested his hand on Alverez's shoulder. "We're going to be fine man, we're going to be just fine."

Lieutenant Colonel Everett sat on the ground with his back leaning against the Humvee that had brought him into El Jasiph. He absentmindedly puffed on a cigar without really realizing it was still in his mouth. His attention was now on a small U.S. flag he had carried in one of his pockets throughout the day's battle. It was now covered in Iraqi sand and had gotten a little bit of blood on it somehow.

Captain Dave Callen walked up to him. "Everything ok, sir? You seem a bit sullen tonight."

"Just thinking, Captain, thanks for your concern," Everett replied.

"Anything you'd like to get off your chest, sir?"

Everett held the flag so Callen could see it. "Just thinking about this is all,"

"The flag, sir?" Callen asked, clearly failing to understand what Everett was telling him.

"Not so much the flag, just how much I never truly understood what the red stripes stood for until today, until El Jasiph."

He was referring to the fact that the red stripes on the American flag stands for the American blood that had to be shed in securing the freedom and liberties Americans enjoy from the British and then defending American freedom in the years since.   Callen nodded his head. "Everything about that flag means more to me today than it did yesterday, sir."

Everett looked up at Callen and smiled weakly.

Lance Corporal Barry Queen's squad manned the roadblock where Charlie Company had entered El Jasiph. The only members of his squad who weren't there were Private First Class Chris Martinez and Private First Class Will Gavan. Both of them were still at the battalion aid station and were more likely to be heading back home because of their injuries.

Private Roger Dole and Queen stood the first watch. They were both on edge and had their M-16s at the ready. It had been a long day and neither knew what to expect on roadblock duty. The rest of the squad did their best to catch a little bit of sleep, but they all failed miserably at it.

The first three hours of the night passed without incident and, with the exception of the explosions coming from inside El Jasiph, was nice and quiet. Then, everything changed. A beat up looking car, that had unquestionably seen better days, started to approach them from El Jasiph. Both, Dole and Queen stood on the road, yelling at the car to stop and motioning for it to stop by displaying their open left palm towards it. Instead of slowing down, the car picked up speed.

"Shit! What do we do now?" Dole asked, his voice on the edge of hysteria.   Queen pulled his M-16 up to his shoulder and Dole followed his example. When the sight of two U.S. Marines with their weapons pointed at it didn't cause the driver to stop, Queen pulled his trigger.

As Queen's bullets struck the windshield and they both saw a sudden flash of red, Dole pulled his trigger. Their bullets shattered the windshield into nothingness, caused the hood to fly up and blew off one of side view mirrors. After several tense seconds, the car veered away from the roadblock and rolled to a stop. By this time the rest of the squad were on their feet and had their weapons pointed at it as well.

"Cover us!" Queen ordered. "Dole, you're with me. You take the driver's side, I'll take the passenger's. You see anything out of the ordinary, you blow anyone that moves in there away. Got me?"

"Got you," Dole replied. Any fear he might have felt had they not spent all day slugging it out with Iraqi soldiers was replaced with a grim determination to carry out his orders. Yesterday he was a Marine, but today he was a combat Marine. There was a difference.

They approached the now motionless car cautiously. When they reached it, they saw two men of military age in civilian clothing. Both were dead. "Reach in and grab the keys out of the ignition," Queen ordered.

Dole looked at him incredulously. "What?"

"We need to clear the trunk and you're on that side. Now get the damn keys."

It was obvious he didn't want to, but Dole reached into the car and pulled the keys from the ignition. Both Marines moved around to the rear of the car. Queen kept his weapon up to his shoulder, ready to fire, while Dole stuck the key into the trunk lock and opened it. Inside were three AK-47s but no ammunition.

"Nothing to do here, back to our positions," Queen said.

Two hours later, Dole had been replaced by Private First Class Leonard Karn. Queen was still in position. Queen knew he wouldn't be able to sleep so he didn't see any reason not to let the Marine who would have taken his place continue to rest. Another car approached from El Jasiph.

"Goddamnit, not again," Queen muttered so softly that Karn hadn't heard him. Just as he had earlier, Queen held out one hand with his palm facing the car and began yelling for it to stop. Karn followed his lead. This time the car slowed down and stopped. Not knowing what to expect, the two Marines approached cautiously. Inside the car was an elderly couple. The husband, who sat behind the steering wheel, motioned forward and said, "Go, go."

Queen shook his head and said, "No go." He motioned back towards El Jasiph and said,

"You go that way."

The husband motioned forward again and said, "Go, go."

"Jesus Christ Almighty, they need to give us translators for this kind of shit," Queen grumbled.

Queen mimed turning the car off. "Give me the keys."

"Go, go," the elderly man said, still motioning forward.

"I'm really getting sick of this bullshit," Queen said.

"Go, go," the old man said, still motioning forward.

Finally having enough of not getting anywhere with the man, Queen opened the driver's side door and gently help him out of his car. The entire time the elderly man kept saying, "Go, go."

Once the husband was out of the car, Queen reached in, turned the car off and removed the keys.

"Get her out of there. This has potential to go downhill on us real fast," Queen said to Karn.

Karn did as he was ordered and plastic flex cuffs were placed on both Iraqis. Queen decided if it turned out the Iraqis weren't up to anything shady, he'd turn their car around for them, unbind their hands, point back towards El Jasiph and say, "Go, go."

Both Marines moved around to the back of the car and opened the trunk. Inside was a middle aged Iraqi man in the uniform of an officer of the Iraqi Army. He held his hands up in surrender.

Queen grabbed the Iraqi officer's arms and pulled him from the trunk while Karn covered him with his M-16. The officer didn't resist and was placed into flex cuffs also.   With all three prisoners secured, Queen got on the Battalion TAC and called for someone to come and pick them up. The rest of their night went peacefully without anyone else trying to run the roadblock.

CHAPTER FIFTY-THREE

The next morning was a hard one. After having had so little sleep on their several day journey from Kuwait to El Jasiph and spending the previous day fighting for their very lives, the Marines were exhausted all the way through to their bones, but they still had work to do.

Their weariness made the camp look more like a scene out of a militarized George A. Romero zombie film than what one would think a Marine camp would look like as the Marines moved about, sluggishly preparing themselves for another day of fighting. The day before, they had experienced real knock-down drag-out fighting in El Jasiph and this day they would be attacking the Iraqi Army base five miles south of the city. No one knew what to expect of the day's fighting. They'd learned the day before not to expect anything in combat because it would likely backfire on you. Very few of them looked forward to the day's fighting—as a matter of fact, most of them dreaded it after what they had already been through. They just hoped whatever lay ahead of them wouldn't be anywhere near as bad for them as the fighting in El Jasiph had been.

Staff Sergeant Dave Ligget and Private Larry Rolling of Bravo Company sat in the backseat of a Humvee, waiting for the orders to roll out to come. In the distance, they could hear explosions coming from the direction of the army base as the Cobra attack helicopters began softening the base for the ground invasion. The helicopters were taking out Iraqi tanks, vehicles of every kind, artillery, anything they could see from the air. There wasn't any way they could completely eliminate every threat to the infantrymen who would be taking control of the base, but they did their best to give the Marines a little less to worry about.

After half an hour, they were on their way and the five miles between El Jasiph and the Iraqi Army base was quickly crossed without incident. Once there, the Marines climbed out of the Humvees and unloaded from the tracks.

The M1-A1 Abrams rolled in first, ready to go nose to nose with any Iraqi tanks that had managed to survive the air assault by the Cobra helicopters.

The battalion's Humvees started to move in next when the sound of mortars being launched reached everyone's ears. They fell wide of the advancing vehicles, which picked up their pace to enter the enemy base before the aim of the Iraqis firing the mortars improved.

As the Marines from all three companies approached on foot, another round of mortars began to fall and the Americans dove to the ground. Once again, the mortars weren't landing anywhere near them, so they got back to their feet and entered the base.

AK-47 bullets began zipping by Sergeant Ed Paige's head as soon as he entered the Iraqi base. "Fuckin' goddamn towelheads!" he yelled, once again living up to his nickname.   Paige saw Iraqi soldiers running around and immediately dropped to one knee. He pulled his M-16 up to his shoulder and started firing at them. He saw three fall but, as far as he was from them, he couldn't tell if they were dead or not. Not that it mattered to him any. After what the Iraqis had put the Marines through the day before, Paige would have been happy to shoot them all in the gut and watch them die painfully. He didn't have anything against the Iraqi civilians, but shooting at him had done nothing to endear the Iraqi soldiers to him any.

He wasn't fired on again immediately, so Paige stood and kept moving forward. The first building he came to was some sort of administration building and Paige entered it with twenty other Marines.

They stormed the administration building as if they expected to find a hundred armed Iraqi soldiers on the other side of the main entrance. They didn't; no one met them. The Marines went about clearing the building quickly, room by room. It was empty.
Most of the Americans left the building, but Paige and four others went to the roof to make sure there wasn't anyone up there. When they kicked open the roof access door, they found two Iraqis firing mortars at the Marines who had remained outside of the base to prevent anyone from escaping and two more firing AK-47s at the Marines who had entered the base and were out in the open.

Everything happened too quickly for the Iraqis to respond. The door opened unexpectedly, and they were all facing away from the Marines who were facing them. Since the Iraqis were all armed and clearly had hostile intent, the Marines didn't give them a chance to surrender.

Paige squeezed his trigger and sent several bullets into the back of one of the Iraqis with an AK-47. The man disappeared over the side of the building. Other Marines fired at the same time and within seconds, all four Iraqis were dead.

With nothing else left to do, all five Marines exited the building and moved on.

Sergeant John Davies and Lance Corporal Kevin Hale found themselves under heavy fire by Iraqis who stood in the open. The two Marines had taken cover behind the burnt out remains of an Iraqi jeep, clearly a casualty of the Cobra attack earlier, and waited for a gap in the incoming enemy fire.

It wasn't cowardice, but the inborn sense for survival all people have that caused them to keep their heads down. The incoming fire was simply too heavy for them to return it, and to try would have been suicide, pure and simple. While neither of them was afraid to die, neither was all that keen on the idea of committing suicide either.

"Shit, Sarge, what're we gonna do?" Hale asked.

Davies didn't answer immediately. He was too busy weighing the pros and cons of several different options he had thought of. Finally, he pulled out a grenade and looked at Hale, saying, "Get ready. As soon as this bad boy goes off, pop up and start firing, don't worry about aiming, just hose them down before anyone who survives the blast has the chance to shoot back at us."
Hale nodded his head in understanding.

Davies pulled the pin on the grenade and let it fly over the top of the jeep in the general direction of where the Iraqis were. It exploded, and the Marines sprang up and started firing their weapons. They both released their triggers after three seconds and all six of the Iraqis who had had them pinned down were dead.

No longer in any immediate danger, Davies looked at Hale and asked, "Are we having fun yet?"

Hale chuckled. "Damn skippy, Sarge. Damn skippy."

It took an hour for the Marines to seize control of the Iraqi Army base and, unlike the day before, it was taken without a single Marine being killed and with only a couple Marines receiving minor wounds that, more than likely, wouldn't end up being enough to end their participation in the war. All in all, it was a good day.

After the day's fighting was done, Lieutenant Colonel Everett watched as the twenty-five captured Iraqi soldiers were loaded onto one of the tracks. A helicopter would pick them up from where the Marines had made camp and take them somewhere— just where, Everett didn't know and didn't care.

What he would have liked to have done was set explosives in every single building on the base and blow them all to pieces—that way there wouldn't be any risk of the Iraqi Army trying to retake it—but his orders were to leave it standing. Someone above him thought that whatever government replaced Saddam's regime would need it.

They still had plenty to do. The next day they would be reentering El Jasiph, which could go either way for them. Everett's gut told him they had pacified the city, that the Iraqi soldiers still alive in El Jasiph had laid down their weapons and were doing their best to blend into the city's civilian population. However, he could be wrong and they could find themselves in another knock-down drag-out fight.

Taking the Iraqi base without losing any more lives had gone a long way toward restoring the morale of his Marines after the bloody battle of El Jasiph. They were visibly happier than they had been the night before and in the morning. He would let them take it easy for the rest of the day, unless they were attacked of course, and his battalion would go back into El Jasiph the next day.

CHAPTER FIFTY-FOUR

The next morning found the Marines in a much better frame of mind than they had been just twenty-four hours earlier. The relatively easy victory at the Iraqi Army base and a rather restful day that followed had done a lot towards helping them be able to relax. When night came, most of them had collapsed into an exhausted and well-earned, but very contented sleep.

A few hours earlier, Colonel Everett had learned that interrogation of the Iraqi officer who had been captured by Lance Corporal Queen's squad while trying to sneak past their roadblock had informed the Marines that the U.S. soldiers from the Army Maintenance convoy that had been captured were being held at a school.

Immediately upon learning that information, he had called a meeting with the battalion command staff and each of the company commanders to develop a rescue plan. The plan was simple: the vast majority of the battalion would enter El Jasiph and start going door to door, looking for Iraqi soldiers, weapons and so on. That would take the enemy's attention off of the school, and two squads would hit it and rescue the American soldiers.

Both squads pulled up to the school in one AAV, they quickly disembarked and took the door of the school. The knowledge that there would be civilians, likely many civilians, and the U.S. soldiers weighed heavily on each of them. They came under fire immediately by Iraqi soldiers who were stationed at the school to guard the prisoners.

Private First Class Jason Campbell squeezed his trigger and sent a small burst of bullets into the chest of one of the Iraqi defenders. The man fell, along with three others, and the rest of the enemy soldiers ran deeper into the school. Whether it was to hide out of cowardliness or to suck the Marines deeper into the school for some strategic reason was anyone's guess, but the Americans advanced.

It took ten minutes for the Marines to find the captured U.S. soldiers. They didn't encounter any more resistance, they didn't find anyone else in the uniform of the Iraqi Army, and they didn't discover weapons of any kind. That being the case, the Marines took the freed American soldiers and left everyone else in the school unbothered.

The silence throughout El Jasiph was deafening as the Marines advanced through the city. Aside from the occasional potshot that someone took at them, none of the Marines came under any enemy fire. The hours that had passed between when they first entered El Jasiph and that point in time had made a world of difference. The surviving Iraqi soldiers, who had managed to avoid being captured, didn't show any interest in continuing the fight. It hadn't been cheap, but the city had been pacified.

Throughout the war, thousands of Americans would pass through El Jasiph on their way deeper into Iraq but none of them would come anywhere near being met with the hostile enemy fire that Alpha, Bravo and Charlie Companies had. The fighting had been tough and intensive but they had accomplished the mission at hand that day, and because of that, hundreds, if not thousands, of lives of the Americans who would need to pass through El Jasiph had been saved. The Marines who fought and died there may never realize it, but they were true American heroes in every sense of the word.

IN HONORED MEMORY

THE MEN AND WOMEN WHO GAVE THEIR LIVES IN AND AROUND AN NASIRIYAH, IRAQ BETWEEN MARCH 22, 2003 AND MARCH 26, 2003. WE WILL NEVER FORGET YOUR SACRIFICE.

1. Sergeant Nicolas M. Hodson, Marine Corps, age 22, Smithville, Missouri

2. SPC Jamaal R. Addison, Army, age 22, Roswell, Georgia

3. Specialist Edward J. Anguiano Army, age 24, Brownsville, Texas

4. Sergeant Michael E. Bitz, Marine Corps, age 31, Ventura, California

5. Lance Corporal Thomas A. Blair, Marine Corps, age 24, Broken Arrow, Oklahoma

6. Lance Corporal Brian Rory Buesing, Marine Corps, age 20, Cedar Key, Florida

7. Sergeant George Edward Buggs, Army, age 31, Barnwell, South Carolina

8. Private First Class Tamario D. Burkett, Marine Corps, age 21, Buffalo, New York

9. Corporal Kemaphoom A. Chanawongse, Marine Corps, age 22, Waterford, Connecticut

10. Lance Corporal Donald "John" Cline Jr., Marine Corps, age 21, Sparks, Navada

11. Master Sergeant Robert J. Dowdy, Army, age 38, Cleveland, Ohio

12. Private Ruben Estrella-Soto, Army, age 18, El Paso, Texas

13. Lance Corporal David K. Fribley, Marine Corps, age 26, Lee, Florida

14. Corporal Jose A. Garibay, Marine Corps, age 21, Costa Mesa, California

15. Private Jonathan L. Gifford, Marine Corps, age 20, Macon, Illinois

16. Corporal Jorge A. Gonzalez, Marine Corps, age 20, Los Angeles, California

17. Private Nolen R. Hutchings, Marine Corps, age 19, Boiling Springs, South Carolina

18. Private First Class Howard Johnson II, Army, age 21, Mobile, Alabama

19. Staff Sergeant Phillip A. Jordan, Marine Corps, age 42, Brazoria, Texas

20. Specialist James M. Kiehl, Army, age 22, Comfort, Texas

21. Chief Warrant Officer 2 Johnny Mata, Army, age 35, El Paso, Texas

22. Corporal Patrick R. Nixon, Marine Corps, age 21, Nashville, Tennessee

23. Private First Class Lori Ann Piestewa, Army, age 23, Tuba City, Arizona

24. Second Lieutenant Frederick E. Pokorney Jr., Marine Corps, age 31, Nye, Nevada

25. Sergeant Brendon C. Reiss, Marine Corps, age 23, Casper, Wyoming

26. Corporal Randal Kent Rosacker, Marine Corps, age 21, San Diego, California

27. Private Brandon Ulysses Sloan, Army, age 19, Bedford, Ohio

28. Lance Corporal Thomas J. Solcum, Marine Corps, age 22, Adams, Colorado

29. Sergeant Donald Ralph Walters, Army, age 33, Salem, Oregon

30. Lance Corporal Michael J. Williams, Marine Corps, age 31, Phoenix, Arizona

31. Private First Class Francisco A. Martinez Flores, Marine Corps, age 21, Los Angeles, California

32. Staff Sergeant Donald C. May Jr., Marine Corps, age 31, Richmond, Virginia

33. Lance Corporal Patrick T. O'Day, Marine Corps, age 20, Sonoma, California

34. Major Kevin G. Nave, Marine Corps, age 36, Union Lake, Michigan

ABOUT THE AUTHORS

JEREMIAH COE

Jeremiah Coe is thirty-five years old and lives in Portage, Michigan. He has been writing since he was five years old and would tell his stories to his mother who would write them down for him.

Jeremiah is the author of four novels, a collection of short stories and four e-books. All four novels are available on Amazon in both paperback and Kindle formats. One of his novels, *Uncivil Dead*, his collection of short stories, Tales From A Twisted Mind Volume One are available for the Kindle and in print on Amazon and in all major e-book formats at Smashwords.com All Four of his e-books are available on Amazon and Smashwords.

He is also the author of several short stories that have been published in various magazines and anthologies.

ADAM D. BLOCHARD

Adam D. Blochard was born in Kalamazoo, Michigan, where he excelled in football and baseball, and received his Eagle Scout badge in Boy Scouts.

He wanted to be a Marine since he was a young boy and left for Marine Corps Recruit Depot San Diego in June of 1995. He has been stationed with various units all over the world and was part of Regimental Combat Team 2 in the 2003 invasion of Iraq. He has served in various jobs including infantry, scout sniper and close combat instructor, as well as with Marine Corps Security Forces. His awards include Navy/Marine Corps Commendation Ribbon with Combat "V" device, Combat Action Ribbon, Presidential Unit Citation, Joint Meritorious Unit Citation, Meritorious Unit Citation, 2 Good Conduct Medals, 2 National Defense Medals, Global War on Terrorism Expeditionary Medal, Global War on Terrorism Service

Medal, 4 Sea Service Deployment Medals, and an Over Seas Service Medal.

ABOUT THEIR FRIENDSHIP

Jeremiah and Adam met at football camp during their freshman year of high school. During high school, they were on the same football and wrestling teams, and were in the Civil Air Patrol, the official Auxiliary of the United States Air Force, together.

They were able to keep in touch after graduating high school in 1995 for a few years, but eventually lost track of each other. Many years later, they reconnected on Facebook and, thanks to the internet; they were able to collaborate on writing Far From Home even though they now live in different states.